Georgy Kitavtsev

Reduced ODE Models Describing Coarsening Dynamics of Liquid Droplets

Georgy Kitavtsev

Reduced ODE Models Describing Coarsening Dynamics of Liquid Droplets

Derivation, Analysis and Numerics of Reduced ODE Models Describing Coarsening Dynamics of Liquid Droplets

Südwestdeutscher Verlag für Hochschulschriften

Impressum / Imprint

Bibliografische Information der Deutschen Nationalbibliothek: Die Deutsche Nationalbibliothek verzeichnet diese Publikation in der Deutschen Nationalbibliografie; detaillierte bibliografische Daten sind im Internet über http://dnb.d-nb.de abrufbar.
Alle in diesem Buch genannten Marken und Produktnamen unterliegen warenzeichen-, marken- oder patentrechtlichem Schutz bzw. sind Warenzeichen oder eingetragene Warenzeichen der jeweiligen Inhaber. Die Wiedergabe von Marken, Produktnamen, Gebrauchsnamen, Handelsnamen, Warenbezeichnungen u.s.w. in diesem Werk berechtigt auch ohne besondere Kennzeichnung nicht zu der Annahme, dass solche Namen im Sinne der Warenzeichen- und Markenschutzgesetzgebung als frei zu betrachten wären und daher von jedermann benutzt werden dürften.

Bibliographic information published by the Deutsche Nationalbibliothek: The Deutsche Nationalbibliothek lists this publication in the Deutsche Nationalbibliografie; detailed bibliographic data are available in the Internet at http://dnb.d-nb.de.
Any brand names and product names mentioned in this book are subject to trademark, brand or patent protection and are trademarks or registered trademarks of their respective holders. The use of brand names, product names, common names, trade names, product descriptions etc. even without a particular marking in this work is in no way to be construed to mean that such names may be regarded as unrestricted in respect of trademark and brand protection legislation and could thus be used by anyone.

Verlag / Publisher:
Südwestdeutscher Verlag für Hochschulschriften
ist ein Imprint der / is a trademark of
OmniScriptum GmbH & Co. KG
Heinrich-Böcking-Str. 6-8, 66121 Saarbrücken, Deutschland / Germany
Email: info@svh-verlag.de

Herstellung: siehe letzte Seite /
Printed at: see last page
ISBN: 978-3-8381-2328-8

Zugl. / Approved by: Berlin, Humboldt University, Diss., 2009

Copyright © 2011 OmniScriptum GmbH & Co. KG
Alle Rechte vorbehalten. / All rights reserved. Saarbrücken 2011

Contents

1	**Introduction**	**3**
	1.1 General Description of the Coarsening Process	3
	1.2 Lubrication Models and their Reduction	5
	1.3 Outline of the Thesis	8
	1.4 Asymptotic Symbols	10
2	**Asymptotical Derivation of Reduced ODE Models**	**11**
	2.1 Stationary Solutions on $\mathbb{R}$	11
	2.1.1 Stationary Solutions for the General Mobility Model	11
	2.1.2 Stationary Solutions for the Strong-slip and Free Films Models	15
	2.2 Near-equilibrium Solutions and Generalized Gradient Flow	15
	2.3 Asymptotical Derivation of Reduced Model for One Droplet	17
	2.3.1 Derivation for the General Mobility Model	17
	2.3.2 Derivation for the Strong-slip and Free Suspended Films Models	19
	2.4 Integration and Asymptotics for Coefficients C_P and C_ξ	22
	2.5 Approximation for the Fluxes between Droplets	26
	2.5.1 Intermediate-slip Case	26
	2.5.2 Strong-slip Case	27
	2.6 Final Form of Reduced ODE Systems	31
	2.7 Numerical Solutions and Comparison	31
	2.7.1 Numerical Methods	31
	2.7.2 Numerical Solutions: Comparison and General Observations	31
	2.8 Numerical Investigation of Inertia Influence	38
3	**Slippage as a Control Parameter for Migration**	**41**
	3.1 Critical Value of Slippage	41
	3.2 Coarsening Patterns for Increasing Slippage	43
	3.3 Coarsening Rates	46
4	**Formal Reduction onto an 'Approximate Invariant' Manifold**	**49**
	4.1 'Approximate Invariant' Manifold: Definition and Properties	50
	4.2 Decomposition in a Neighborhood of the Manifold	58
	4.3 Formal Leading Order for Equation on the Manifold	60
	4.4 Comparison of Reduced ODE Models	63
	4.5 Discussion and Spectral Problem	63
5	**Spectrum Asymptotics in a Singular Limit**	**65**
	5.1 Scalings and Linearized Eigenvalue Problems	65
	5.2 Summary of Main Results and Discussion	69
	5.3 Half-droplet Problem and its Approximations	72
	5.4 Asymptotics for Stationary Solutions	76
	5.5 Spectrum Asymptotics for the Approximate Problems	84
	5.6 Existence of Eigenvalues with Prescribed Asymptotics	92

Contents

- 5.7 Proof of the Main Theorems 100
- 5.8 Numerical Solutions and Comparison 104

6 Summary and Outlook   **107**

List of main symbols   **109**

Bibliography   **111**

Chapter 1
Introduction

1.1 General Description of the Coarsening Process

The last several decades showed considerable interest and intensive research among scientists and engineers on topics concerning such physical processes and phenomena as dewetting in micro and nanoscopic liquid films on a solid polymer substrate. There is a large number of applications of dewetting processes in several brunches of physics, chemistry and material sciences. Among them are the evolution of free liquid surface during coating and printing processes, see e.g. Oron et al. [6], development of Lab-on-chip devices and liquid crystal displays, see e.g. Granick et al. [7], Jacobs et al. [8]. In general, such dewetting processes can be divided into three stages.

During the first stage a liquid polymer film of nanometer thickness interacting with a hydrophobically coated solid substrate is susceptible to instability due to small perturbations of the film profile. Typically such films rupture, thereby initiating a complex dewetting process, see e.g. Reiter et al. [9], Redon et al. [10], Seemann et al. [11]. The influence of intermolecular forces play an important part in the rupture and subsequent dewetting process, see e.g. Oron et al. [6], de Gennes [12], Williams and Davis [13] and references therein. Typically the competition between the long-range attractive van der Waals and short-range Born repulsive intermolecular forces reduces the unstable film to *an ultra-thin layer* that connects the evolving patterns and is given by the minimum of the corresponding intermolecular potential, i.e. the film settles into an energetically more favorable state, see Erneux and Gallez [14], Bertozzi et al. [15]. The second stage is associated with the formation of regions of this minimal thickness, bounded by moving rims that connect to the undisturbed film, see e.g. Sharma and Reiter [16], Brochard-Wyart and Redon [17], Münch and Wagner [18].

In this study we are interested in the third and the last stage of the dewetting process, namely the long-time coarsening process that originates in the breaking up of the evolving patterns into small droplets and is characterized by its subsequent slow-time coarsening dynamics, which has been observed and investigated experimentally by Limary and Green [19, 20]. In the mathematical modeling of thin films it has been shown to be of great advantage to reduce the governing equations to an equation for the free surface $h(x,t)$ using lubrication theory (see a geometrical sketch on Figure 1.1). Experimental observations from Limary and Green [19] of the coarsening process in three-dimensional case show that during the coarsening the average size of droplets increases and the number of droplets decreases. The coarsening mechanisms that were observed in such films are typically collapse of the smallest droplets and collision of neighboring ones. During collapse the size of a droplet shrinks in time and its mass is distributed in the ultra-thin layer. Collisions among droplets occur due to the mass transfer through the ultra-thin layer between them that causes a translation movement of them, *droplet migration*, eventually leading to the formation of new droplets. A numerical example of the coarsening dynamics in two-dimensional case is shown in Figure 2.2. Besides intermolecular forces and surface tension at the free surface it has been shown by Fetzer et al. [21] that the dewetting of polymer films on hydrophobic substrates also involves such boundary effect as slippage on a solid substrate. The measure of slip is a so-called slip length, which is defined as an extrapolated distance relative to the wall where

Chapter 1 Introduction

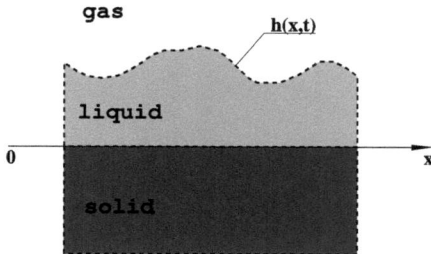

Figure 1.1: Geometrical sketch for a two-dimensional liquid film on a solid substrate with the free surface described by function $h(x,t)$.

the tangential velocity component u vanishes, see Figure 1.2. A commonly used expression for it is given by the Navie-slip boundary condition

$$b := \frac{u_z}{u}, \qquad (1.1)$$

where u_z is the derivative of u in the direction normal to the solid substrate. As it is illustrated in Figure 1.2 the slip length determines physically a type of the velocity profile in the liquid. In the no-slip case $b = 0$ it is assumed that the liquid does not move at the contact points with the solid surface and a typical flow is parabolic. In another limiting case $b = \infty$ the tangential velocity does not change in the normal direction to the solid substrate (plug flow profile). For finite values of b the velocity profile with partial slip changes continuously between above limiting ones. Recently, it has been shown experimentally and theoretically that the early stages of

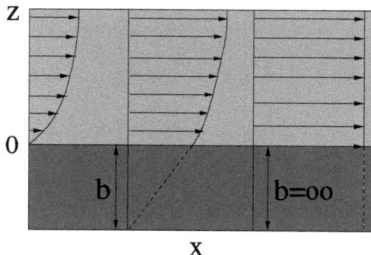

Figure 1.2: Three different flow profiles corresponding to the no-slip situation (left), partial-slip case with a finite slip length (middle) and plug flow (right), when the effective slip length becomes infinite.

the dewetting process and the evolving morphology depend markedly on the magnitude of the effective slip length, which can be of the size of the height of the liquid film or even larger for nanoscale systems, see e.g. Münch and Wagner [18], Neto et al. [22], Redon et al. [23], Reiter and Sharma [24], Fetzer et al. [25]. As was stated in Münch et al. [1] the order of magnitude of the effective slip length value influences the choice of an appropriate mathematical model describing the coarsening process. One of the aims of this study is an analysis of the influence of slippage on the late stage long-time coarsening process.

1.2 Lubrication Models and their Reduction

Throughout our study we deal with two-dimensional films on a one-dimensional solid substrate as in Figure 1.1. The general mathematical model describing the evolution of such films is given by the two-dimensional Navier-Stokes equations coupled with conservation of mass together with appropriate boundary conditions on the solid and free surface. As was mentioned in the previous section the complicated behavior of liquid films is conditioned by such physical effects as surface tension at the free boundary, intermolecular interactions with solid substrate and an effective slip length on the latter one. In order to understand this behavior, and using an obvious vertical to horizontal scale separation in such films, in Münch et al. [1], Kargupta et al. [26] closed-form one-dimensional lubrication models over a wide range of slip lengths were derived from the underlying equations for conservation of mass and momentum, together with boundary conditions for the tangential and normal stress, as well as the kinematic condition at the free boundary, impermeability and Navier-slip condition at the liquid-solid interface. Asymptotic arguments, based on the magnitude of the slip length show that within a lubrication scaling there are two *distinguished limits*, see Münch et al. [1].

These are the well-known *weak-slip* model

$$\partial_t h = -\partial_x \Big((h^3 + b\, h^2) \partial_x \left(\partial_{xx} h - \Pi_\varepsilon(h) \right) \Big) \tag{1.2}$$

with b denoting the slip-length parameter, and the *strong-slip* model

$$Re\left(\partial_t u + u \partial_x u\right) = \frac{4}{h} \partial_x (h \partial_x u) + \partial_x \left(\partial_{xx} h - \Pi_\varepsilon(h)\right) - \frac{u}{\beta h}, \tag{1.3a}$$

$$\partial_t h = -\partial_x (hu), \tag{1.3b}$$

respectively. Here, $u(x,t)$, $h(x,t)$ denote the average velocity in the lateral direction and height profile for the free surface, respectively. The slip-length parameters b and β are related by orders of magnitude via $b \sim \nu^2 \beta$, where the parameter ν with $0 < \nu \ll 1$ refers to the vertical to horizontal scale separation of the thin film. The high order of the lubrication equations (1.2) and (1.3a)–(1.3b) is a result of the contribution from surface tension at the free boundary, reflected by the linearized curvature term $\partial_{xx} h$. A further contribution to the pressure is denoted by $\Pi_\varepsilon(h)$ and represents one from the intermolecular forces, namely long-range attractive van der Waals and short-range Born repulsive intermolecular forces. A commonly used expression for it is given by

$$\Pi_\varepsilon(h) = \frac{\varepsilon^2}{h^3} - \frac{\varepsilon^3}{h^4}, \tag{1.4}$$

It can be written as a derivative of the potential function $U_\varepsilon(h)$,

$$U_\varepsilon(h) = -\frac{\varepsilon^2}{2\, h^2} + \frac{\varepsilon^3}{3\, h^3}, \tag{1.5}$$

where parameter $0 < \varepsilon \ll 1$ is the global minimum of the latter function and gives to the leading order thickness of the ultra-thin layer (see Figure 1.3). The terms $Re\left(\partial_t u + u \partial_x u\right)$, with Re denoting the Reynolds number, and $(4/h)\partial_x(h\, \partial_x u)$ in (1.3a)–(1.3b) are called the inertial and Trouton viscosity terms, respectively.

Additionally, the weak-slip and the strong-slip models contain as limiting cases three further lubrication models. One of them is the *no-slip model*, which is obtained setting $b = 0$ in the weak-slip model:

$$\partial_t h = -\partial_x \Big(h^3 \partial_x \left(\partial_{xx} h - \Pi_\varepsilon(h) \right) \Big). \tag{1.6}$$

Chapter 1 Introduction

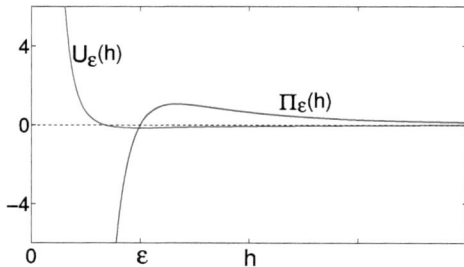

Figure 1.3: Plots of intermolecular pressure $\Pi_\varepsilon(h)$ (blue) and potential function $U_\varepsilon(h)$ (green) for $\varepsilon = 0.1$

The second one is obtained from the strong-slip model in the limit $\beta \to \infty$ and describes the dynamics of suspended free films, see e.g. Brenner and Gueyffier [27]:

$$Re\left(\partial_t u + u \partial_x u\right) = \frac{4}{h}\partial_x(h\partial_x u) + \partial_x\left(\partial_{xx} h - \Pi_\varepsilon(h)\right), \tag{1.7a}$$

$$\partial_t h = -\partial_x(hu), \tag{1.7b}$$

For the third limiting case derived in Münch et al. [1] the slip-length parameter β_I is of order of magnitude lying in between those that lead to the weak and the strong-slip model, i.e. $b \ll \beta_I \ll \beta$. The corresponding *intermediate-slip* model is given by

$$\partial_t h = -\partial_x\left(h^2 \partial_x\left(\partial_{xx} h - \Pi_\varepsilon(h)\right)\right). \tag{1.8}$$

It can be obtained by rescaling time in (1.2) by b and letting $b \to \infty$ or by rescaling time and the horizontal velocity by β in (1.3a)–(1.3b) and taking the limit $\beta \to 0$.

The no-slip, weak-slip and intermediate slip lubrication models are given by a parabolic equations for the height profile $h(x,t)$, which degenerates as $h \to 0$. For convenience of our analytical investigation of these models we write them below in a general form, which we call *general mobility model*:

$$\partial_t h = -\partial_x\left(M(h)\partial_x\left(\partial_{xx} h - \Pi_\varepsilon(h)\right)\right). \tag{1.9}$$

This equation incorporates the three former lubrication models for particular cases of the nonlinear mobility term $M(h)$. For example in the no-slip case $M(h) = h^3$. In this study we often describe (1.9) on a bounded interval $(-L, L)$ with boundary conditions

$$\partial_{xxx} h = 0, \quad \text{and} \quad \partial_x h = 0 \quad \text{at} \quad x = \pm L, \tag{1.10}$$

which incorporate zero flux at the boundary and as a consequence imply the conservation of mass law:

$$h_c = \frac{1}{2L}\int_{-L}^{L} h(x,t)\,dx, \; \forall t > 0, \tag{1.11}$$

where $h_c = \text{const}$ is the average of the height profile. It has been shown by Bertozzi et al. [15] that the general mobility model (1.9) with boundary conditions (1.10) and initial data $h_0(x)$ has a unique strong positive solution, provided that $h_0(x) \in H^1(-L, L)$, positive for all $x \in (-L, L)$

and
$$\int_{-L}^{L} \frac{1}{2}|\partial_x h_0(x)|^2 + U_\varepsilon(h_0(x))\,dx < \infty,$$
where $U_\varepsilon(h)$ is defined in (1.5).

In this study we consider systems (1.3a)–(1.3b) and (1.7a)–(1.7b) on interval $(-L, L)$ with the following boundary conditions. We put velocities (or fluxes) at the boundary to zero

$$u = 0 \quad \text{at} \quad x = \pm L. \tag{1.12}$$

i.e. we require conservation of mass (1.11). For the profile $h(x, t)$ we assume that

$$\partial_x h = 0 \quad \text{at} \quad x = \pm L. \tag{1.13}$$

The same boundary conditions were used by Peschka [28] for analytical and numerical investigation of rupture processes driven by (1.3a)–(1.3b).

Within the context of thin liquid films one of the first studies of the coarsening dynamics can be found in Glasner and Witelski [2] and Glasner and Witelski [29]. These authors consider the one-dimensional no-slip lubrication model (1.6) with (1.10). They confirmed numerically the existence of two coarsening driven mechanisms discussed in the previous section, namely collision and collapse. Our numerical investigations (see section 2.7) of particular cases of the model (1.9) and of the strong-slip model (1.3a)–(1.3b) with sufficiently small Re number all identified also that the coarsening dynamics is driven by these two coarsening mechanisms. Nevertheless, in applications the number of droplets can be very large, of order 10^3. For example, one of the typical problems considered also in this study is the calculation of the coarsening rates, i.e. how fast the number of droplets decreases during the coarsening dynamics depending on different physical parameters. Often in order to identify the characteristic dependence for coarsening rates one needs to model very large arrays of droplets. But due to the presence of the ultrathin-layer with order ε between each pair of droplets the problem of numerical solution for any lubrication model becomes very stiff in time and demands high space resolution as the number of droplets increases. Therefore, in the mathematical modeling of such thin films there exists a need for further reduction of lubrication models to more simple, possibly finite-dimensional ones.

Within a different context of phase separation of binary alloys, coarsening dynamics is a well-known widely studied process and is typically described by the Cahn-Hilliard equation, see e.g. Cahn and Hilliard [30]. For the late phases of this process the existence of near-equilibrium solutions was first shown by Alikakos et al. [31]. After that reduced ODE models have been derived and investigated by Bates and Xun [32, 33], San and Ward [34] that allow to determine properties such as coarsening rates, which can be time consuming using the underlying partial differential equations. These studies have recently been extended to describe phase separation under the influence of an external driving field by Emmott and Bray [35], Watson et al. [36]. The driven mechanisms in this case is given by Ostwald ripening. As was recently pointed out by Glasner et al. [3] in the case of thin liquid films the collapse component of the coarsening dynamics is analogous to Ostwald ripening in binary alloys. But the additional coarsening component in thin films, namely collisions and migration effect of droplets, makes the dynamics in some sense richer.

In Glasner and Witelski [2] and Glasner and Witelski [29] for the first time a reduced ODE model was derived from the lubrication no-slip equation. This model was used for an effective analysis of the coarsening rates. Additionally, using essentially a mixture of the gradient flow structure approach and asymptotic analysis, reduced ODE models for the one as well as two-dimensional case for the general mobility model with $M(h) = h^q$, $q > 0$ were recently derived in Glasner et al. [3]. Within the different context of Darcy's equation for the case $M(h) = h$

Chapter 1 Introduction

coarsening rates on the basis of the gradient flow structure for the corresponding equation were derived by Otto et al. [37]. Moreover, for this case they showed that the analysis can be made rigorous. One focus of the work of Glasner et al. [3] concerned migration and its underlying causes, where results of Pismen and Pomeau. [38] on the relation of the direction of the droplet motion and mass flux were discussed and clarified, i.e. that, indeed, in the systems governed by a type of the general mobility model the direction of the migration of droplets is opposite to the applied mass flux. In a recent paper by Glasner [39] results of Glasner and Witelski [2] were extended to the two-dimensional general mobility model with $M(h) = h^q$, $q > 0$ and comparison with the alternative derivation from Glasner et al. [3] was given.

In view of the above developments, in this study we consider several new questions concerning the coarsening dynamics in thin liquid films. The general aim is to generalize previous results in the one dimensional case and derive a complete set of reduced ODE models for all lubrication models stated above. Next, we would like to understand the effects of slippage on details of the coarsening mechanisms via the reduced ODE models. Finally, we would like to look at the methods for rigorous mathematical justification of such reduced models, the question to which up to date there is no a complete answer.

1.3 Outline of the Thesis

In this thesis the topic of derivation, analysis and numerics of reduced ODE models corresponding to the set of lubrication equations stated in the previous section is addressed. Besides their asymptotical derivation we give an analytical and numerical investigation of these models. In particular, the influence of slippage on the coarsening dynamics is analyzed via the reduced ODE models. Some new methods for the rigorous justification of these models are suggested.

For this purpose the thesis is divided in six parts. In **Chapter 2** we begin with the derivation of special type of positive stationary solutions for lubrication equations considered on the whole real line $\mathbb{R}$. In section 2.3 using formal asymptotical analysis we derive a reduced finite dimensional models for the general mobility model (1.9) (covering by that weak, no- and intermediate-slip cases) and for the strong-slip model (1.3a)–(1.3b) with sufficiently small Re number. The reduced ODE models describe the effective dynamics of droplets in an array and govern the evolution of their pressures and positions in time. It turns out that in the case of the general mobility model (1.9) the derivation of the corresponding reduced ODE model can be obtained by following the ideas of Glasner and Witelski [2], so that we only briefly summarize our results here and focus mainly on the strong-slip model in paragraph 2.3.2. In section 2.5 to make reduced ODE models complete we derive asymptotically approximations for the fluxes between droplets in an array. In comparison with results of Glasner and Witelski [2] the new here are approximations for the intermediate and strong-slip case. Finally, in section 2.7 we present numerical schemes for the solution of lubrication equations and corresponding reduced ODE models. Whereas our numerical scheme for lubrication equations is based on the one developed by Münch et al. [1], Münch [40] and Peschka [28], new schemes are constructed for the integration of reduced ODE models and applied later to numerical simulations of coarsening rates. In section 2.7 we also compare numerical results for the lubrication equations and the corresponding ODE reduced models and give general observations concerning validity and properties of the latter ones. We conclude the second chapter with a preliminary numerical analysis for the coarsening dynamics governed by the strong-slip model (1.3a)-(1.3b) with a moderate Re numbers. We identify here new interesting coarsening effects in comparison with already known for the case of a small Re number.

In **Chapter 3** analyzing the reduced model for the strong-slip model (1.3a)–(1.3b) we show that in contrast to the general mobility model, which was treated by Glasner et al. [3] with mobility $M(h) = h^q$, $q > 0$, in the strong-slip case a droplet does not necessarily migrate in the direction opposite to the applied mass flux. There is a critical value of the slippage $\beta = \beta_{crit}$ such that for slip-lengths bigger then β_{crit} droplets migrate in the direction of the flux. As a further consequence of that we find that collisions of two droplets are possible for some range of slip parameter β in the equation (1.3a)-(1.3b), while for the cases described by (1.6), (1.2) and (1.8) as was shown by Glasner et al. [3], Glasner and Witelski [29] collisions involve at least three droplets. In section 3.2 we investigate numerically using derived reduced ODE models the resulting coarsening patterns with increasing slippage. Simulating numerically large arrays of droplets we identify another new effect for the strong-slip case. We observe that, due to the existence of β_{crit} changing the slip length β influences considerably the coarsening scenario and relative proportion of coarsening events (collapse or collision). For the general mobility model (1.9) with $M(h) = h^q$, $q > 0$, it was shown in Glasner et al. [3] that the collision component of the coarsening process is negligible in comparison with the collapse component for $q < 3$ and becomes comparable only in the case $q = 3$. In contrast to that we observe in the strong-slip case that the collision component increases, when β increases starting from $\beta = \beta_{crit}$, and becomes the dominant mechanism of the coarsening process. Finally, using reduced ODE models we carry out numerical simulations of coarsening rates in the strong-slip case and analyze the influence of slippage on them.

In **Chapter 4** we give an alternative formal derivation of the reduced ODE model corresponding to the no-slip equation (1.6). This derivation is motivated by a recent article of Mielke and Zelik [4] where a center invariant manifold approach was applied to a rather general type of semilinear parabolic equations in order to obtain reduced ODE systems for them. Following formally this approach we end up finally with an alternative reduced ODE model. Our approach is based on two steps. In section 4.1 we construct a so called 'approximate invariant' manifold parameterized by a set of positions and pressures in a droplet array. In the next two sections we do a formal reduction of (1.6) onto this manifold and derive an alternative reduced ODE model. We compare it in section 4.4 with the one derived asymptotically in Chapter 1 and by Glasner and Witelski [2] for the no-slip case and find a good agreement between them. Nevertheless, the rigorous justification of a center-manifold reduction in the case of the no-slip equation is a more complicated problem than those described by Mielke and Zelik [4], because (1.6) is a quasilinear equation, which additionally degenerates as $h \to 0$. In section 4.5 we discuss more precisely some open questions that are needed to be solved for justification of the above approach and formulate a so called spectral problem.

Motivated by this problem in **Chapter 5** we derive rigorously the asymptotics for the spectrum of the no-slip model (1.6) linearized at a stationary droplet solution in the limit $\varepsilon \to 0$, where the small parameter $\varepsilon > 0$ appears in all lubrication models through the intermolecular pressure function (1.4). It turns out that the resulting linear eigenvalue problem (5.12) is singularly perturbed as $\varepsilon \to 0$. The main results on its spectrum asymptotics are given in Theorems 5.10-5.12. They state that in the spectrum of the above linear eigenvalue problem there exists a set of algebraically small eigenvalues and an exponentially small one as $\varepsilon \to 0$. Between the former set and the latter eigenvalue there exists an ε-dependent spectral gap. The main ingredients for proving Theorem 5.10-5.12 in sections 5.3–5.7 are approximate eigenvalue problems and the modified implicit function Theorem 5.30, first introduced by Magnus [41] and Recke and Omel'chenko [5]. We conclude this chapter with a numerical solution of the linearized eigenvalue problem (5.12) and a comparison of it with the analytical results of Theorems 5.10-5.12. In **Chapter 6** the summary and outline for the thesis are stated.

Chapter 1 Introduction

1.4 Asymptotic Symbols

In this study we often describe asymptotical processes with respect to the small parameter ε introduced first in (1.4). Many of the functions in the text depend on it. In order to escape from any ambiguity in the treatment of asymptotical processes and the corresponding symbols we give below the definition for the latter ones, which is used everywhere in this text. Note that it corresponds to the definition of the asymptotical symbols from section 1.1 of Erdelyi [42] in the case $\varepsilon \to 0$ and holding *uniformly* in a parameter set D.

Definition 1.1. Let $\varepsilon_0 > 0$, functions $f, g : (0, \varepsilon_0) \times \mathbb{R}^m \to \mathbb{R}$ with $m \geq 0$ and a set $D \subset \mathbb{R}^m$ be given.

(i) We write $f = O(g)$ for all x in D if and only if there exist numbers $M > 0$ and $\varepsilon_1 \in (0, \varepsilon_0)$ such that
$$|f(\varepsilon, x)| \leq M |g(\varepsilon, x)| \text{ for all } x \in D \text{ and } \varepsilon \in (0, \varepsilon_1). \tag{1.14}$$

(ii) We write $f = o(g)$ for all x in D if and only if for any given $\delta > 0$ there exists $\varepsilon_1(\delta) \in (0, \varepsilon_0)$ such that
$$|f(\varepsilon, x)| \leq \delta |g(\varepsilon, x)| \text{ for all } x \in D \text{ and } \varepsilon \in (0, \varepsilon_1). \tag{1.15}$$

(iii) We write $f \sim g$ for all x in D if and only if $f - g = o(g)$.

In Appendix we collect a list of other main symbols and notation used in this text.

Chapter 2
Asymptotical Derivation of Reduced ODE Models

2.1 Stationary Solutions on $\mathbb{R}$

2.1.1 Stationary Solutions for the General Mobility Model

In Bertozzi et al. [15], Glasner and Witelski [29] a special kind of positive stationary solutions to the no-slip model (1.6) was described. For our subsequent analysis it will be useful to generalize their results for the general mobility model (1.9) in the following theorem.

Theorem 2.1. *Equation (1.9) considered on the whole real line $\mathbb{R}$ has a family of positive nonconstant steady state solutions $\hat{h}_\varepsilon(x, P)$ parameterized by a constant (a so called pressure) $P \in (0, P_{max}(\varepsilon))$, where*

$$P_{max}(\varepsilon) := \frac{27}{256\varepsilon}, \qquad (2.1)$$

which satisfy

$$\partial_{xx}\hat{h}_\varepsilon(x, P) = \Pi_\varepsilon(\hat{h}_\varepsilon(x, P)) - P, \qquad (2.2a)$$

$$\hat{h}_\varepsilon(x, P) = \hat{h}_\varepsilon(-x, P), \qquad (2.2b)$$

$$\partial_x \hat{h}_\varepsilon(0, P) = 0 \quad and \quad \partial_x \hat{h}_\varepsilon(x, P) < 0 \text{ for } x > 0. \qquad (2.2c)$$

For fixed positive numbers $P^* > P_* > 0$ the following asymptotics holds for all $P \in (P_*, P^*)$:

$$\hat{h}_\varepsilon^-(P) := \min_{x \in \mathbb{R}} \hat{h}_\varepsilon(x, P) = \epsilon + \epsilon^2 P + O(\epsilon^3). \qquad (2.3a)$$

$$\hat{h}_\varepsilon^+(P) := \max_{x \in \mathbb{R}} \hat{h}_\varepsilon(x, P) = \frac{1}{6P} + O(\epsilon). \qquad (2.3b)$$

Proof: For each $\varepsilon > 0$ it is simple to deduce that any solution to equation

$$h'' = \Pi_\varepsilon(h) - P, \qquad (2.4)$$

with P being a number gives a stationary solution to (1.9) on $\mathbb{R}$. The rest of the proof can be done via a phase plane analysis for equation (2.4) described in Bertozzi et al. [15] (see also Figure 2.1). It shows that for any fixed $P \in (0, P_{max}(\varepsilon))$ there exists a homoclinic loop $\hat{h}_\varepsilon(x, P)$ for equation (2.4). The value (2.1) for $P_{max}(\varepsilon)$ is given by the global maximum of $\Pi_\varepsilon(h)$, which is attained at $h_{max} = 4/3\varepsilon$. Moreover, there exists a phase shift such that $\hat{h}_\varepsilon(x, P)$ satisfy also (2.2b)–(2.2c). The asymptotics (2.3a)–(2.3b) were derived in Glasner and Witelski [29]. The smallest real root of algebraic equation $\Pi_\varepsilon(h) = P$ is a saddle-point to equation (2.4) and gives us $\hat{h}_\varepsilon^-(P)$. Expanding identity $\Pi_\varepsilon(\hat{h}_\varepsilon^-(P)) = P$ in ε one obtains (2.3a). An elliptic center point $\hat{h}_\varepsilon^c(P)$ of equation (2.4) is the other real root of $\Pi_\varepsilon(h) = P$ and has asymptotics:

$$\hat{h}_\varepsilon^c(P) = \epsilon(\epsilon P + o(\epsilon))^{-1/3}. \qquad (2.5)$$

Chapter 2 Asymptotical Derivation of Reduced ODE Models

Once $\hat{h}_\varepsilon^-(P)$ is determined, the first integral to equation (2.2a) can be written as

$$\frac{1}{2}\left(\partial_x \hat{h}_\varepsilon(x,\,P)\right)^2 + \mathcal{U}_\varepsilon(\hat{h}_\varepsilon(x,\,P),\,P) = 0, \tag{2.6}$$

where

$$\mathcal{U}_\varepsilon(h,\,P) := -U_\varepsilon(h) + U_\varepsilon(\hat{h}_\varepsilon^-(P)) + P(h - \hat{h}_\varepsilon^-(P)). \tag{2.7}$$

By (2.2b)–(2.2c) $\hat{h}_\varepsilon(x,\,P)$ attains its maximum at $x = 0$, and therefore $\hat{h}_\varepsilon^+(P)$ is determined by the condition $\mathcal{U}_\varepsilon\left(\hat{h}_\varepsilon^+(P),\,P\right) = 0$. Again, after expansion of the last identity in ε one obtains (2.3b). ∎

More detailed asymptotic analysis of $\hat{h}_\varepsilon(x, P)$ as $\varepsilon \to 0$ (see Glasner [39]) shows that it can be described by a parabola connected to a thin layer of order ε and looks like *a droplet*. $\hat{h}_\varepsilon^-(P)$ gives to the leading order in ε the thickness of the thin layer and $\hat{h}_\varepsilon^+(P)$ the peak of the droplet (see Figure 2.1).

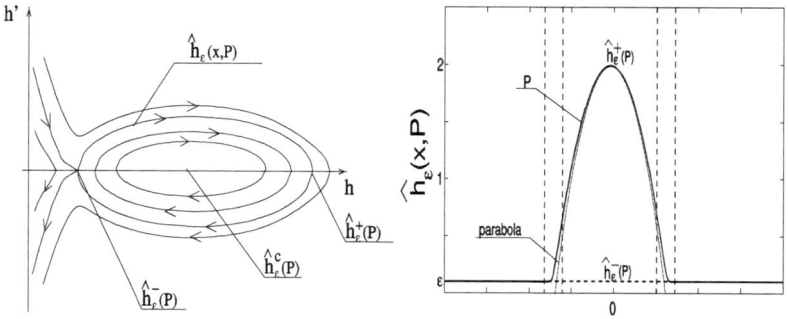

Figure 2.1: Phase plane portrait for the equation (2.4) (left) and plot of stationary solution $\hat{h}_\varepsilon(x,\,P)$ (right).

The next proposition states additional asymptotic properties for $\hat{h}_\varepsilon(x,\,P)$, which are used during derivation of reduced ODE models in sections 2.3 and 4.1.

Proposition 2.2. *There exist positive numbers d, $P^* > P_*$ and C_k, $k = 0, 1$ such that for all $|x| > d$, $P \in (P_*,\,P^*)$ and sufficiently small $\varepsilon > 0$ one has*

$$\left|\hat{h}_\varepsilon(x,\,P) - \hat{h}_\varepsilon^-(P)\right| \leq C_0 \exp\left(\frac{d-x}{\sqrt{2}\varepsilon}\right), \tag{2.8a}$$

$$\left|\frac{\partial^k \hat{h}_\varepsilon(x,\,P)}{\partial x^k}\right| \leq \frac{C_0}{\varepsilon^k} \exp\left(\frac{d-x}{\sqrt{2}\varepsilon}\right) \text{ for } k = 1, 2, 3, 4, \tag{2.8b}$$

$$\frac{\partial \hat{h}_\varepsilon(x,\,P)}{\partial P} \leq C_1 \varepsilon (x - d). \tag{2.8c}$$

Proof: Let us define a function

$$F(v) := -\mathcal{U}_\varepsilon(v + \hat{h}_\varepsilon^-(P),\,P),$$

where $\mathcal{U}_\varepsilon(h,\,P)$ is defined by (2.7). From the proof of Theorem 2.1 it follows that

2.1 Stationary Solutions on $\mathbb{R}$

$\Pi(\hat{h}_\varepsilon^-(P)) - P = 0$. Using this and (2.7) one obtains

$$F(0) = -\mathcal{U}_\varepsilon(\hat{h}_\varepsilon^-(P), P) = 0,$$
$$F'(0) = 0,$$
$$F''(v) = \Pi'_\varepsilon(v + \hat{h}_\varepsilon^-(P)).$$

Therefore, applying Newton-Leibniz formula to $F(v)$ and integrating once by parts one gets

$$F(v) = \int_0^1 (1-t)\Pi'_\varepsilon(t(v + \hat{h}_\varepsilon^-(P)))\, dt\, v^2$$

Substituting in the last expression $v_\varepsilon(x, P) := \hat{h}_\varepsilon(x, P) - \hat{h}_\varepsilon^-(P)$ and using (2.6), (2.2c) one obtains that

$$\frac{\partial_x v_\varepsilon(x, P)}{v_\varepsilon(x, P)} = -\sqrt{2\left(\int_0^1 (1-t)\Pi'_\varepsilon\left(t\hat{h}_\varepsilon(x, P)\right) dt\right)} \quad \text{for } x > 0. \tag{2.9}$$

By (1.4) and (2.3a) function $\Pi'_\varepsilon(h)$ monotonically decays on $[\hat{h}_\varepsilon^-(P), 4/3\,\varepsilon]$ to zero and

$$\Pi'_\varepsilon(\hat{h}_\varepsilon^-(P)) \sim 1/\varepsilon^2. \tag{2.10}$$

Using this and (2.2c) let us define uniquely $\nu_\varepsilon(P) > 0$ such that

$$\Pi'_\varepsilon\left(\hat{h}_\varepsilon(\nu_\varepsilon(P), P)\right) := \frac{1}{2\varepsilon^2} \tag{2.11}$$

Next, we fix some positive numbers $P^* > P_* > 0$ and show using a contradiction argument that there exists a number $d > 0$ such that $d > \nu_\varepsilon(P)$ for all sufficiently small $\varepsilon > 0$ and $P \in (P_*, P^*)$. Suppose inverse then there should exist sequences $\{P_n\}$, $\{\varepsilon_n\}$ with $P_n \in (P_*, P^*)$ for all $n \in \mathbb{N}$ and $\varepsilon_n \to 0$ such that $\nu_{\varepsilon_n}(P_n) \to +\infty$ as $n \to +\infty$. Using asymptotics (2.3a), (2.5) and (2.2c) one obtains that there exists a positive number $\tilde{\varepsilon}$ such that

$$\hat{h}_\varepsilon(x, P) - \hat{h}_\varepsilon^-(P) \to 0 \text{ as } x \to \infty \text{ uniformly in } \varepsilon \in (0, \tilde{\varepsilon}) \text{ and } P \in (P_*, P^*),$$

and hence using (2.10) one concludes

$$\frac{\Pi'_{\varepsilon_n}\left(\hat{h}_{\varepsilon_n}(\nu_{\varepsilon_n}(P_n), P_n)\right)}{1/\varepsilon_n^2} \to 1 \text{ as } n \to \infty.$$

But the last expression gives a contradiction to definition (2.11). Therefore, number $d > 0$ with above properties exists.

Let us now fix any $x > d$. Using monotonicity of $\Pi'_\varepsilon(h)$, (2.2c), (2.3a) and definition of d one obtains

$$\frac{1}{2\varepsilon^2} < \Pi'_\varepsilon(\hat{h}_\varepsilon(d, P)) \leq \Pi'_\varepsilon(\hat{h}_\varepsilon(x, P)) < \Pi'_\varepsilon(\hat{h}_\varepsilon^-) \leq \frac{1}{\varepsilon^2} \tag{2.12}$$

for sufficiently small $\varepsilon > 0$ and $P \in (P_*, P^*)$. Integrating (2.9) on $(\nu_\varepsilon(P), x)$ and using (2.12) one estimates

$$\frac{v_\varepsilon(x, P)}{v_\varepsilon(\nu_\varepsilon(P), P)} = \exp\left(-\int_{\nu_\varepsilon(P)}^x \sqrt{2\int_0^1 (1-t)\Pi'_\varepsilon(t\,\hat{h}_\varepsilon(x, P))\,dt}\, dx\right)$$
$$\leq \exp\left[\frac{d-x}{\sqrt{2}\varepsilon}\right].$$

Chapter 2 Asymptotical Derivation of Reduced ODE Models

From (2.11) and definition of $v_\varepsilon(x, P)$ it follows that
$$v_\varepsilon(\nu_\varepsilon(P), P) \leq \hat{h}_\varepsilon(\nu_\varepsilon(P), P) \leq C_0,$$
where constant C_0 does not depend on ε and P, and therefore
$$\left|\hat{h}_\varepsilon(x, P) - \hat{h}_\varepsilon^-(P)\right| \leq C_0 \exp\left[\frac{d-x}{\sqrt{2}\varepsilon}\right]. \tag{2.13}$$

Next, by (2.9) and (2.12) one obtains
$$\left|\partial_x \hat{h}_\varepsilon(x, P)\right| \leq \frac{1}{\varepsilon}\left|\hat{h}_\varepsilon(x, P) - \hat{h}_\varepsilon^-(P)\right| \leq \frac{C_0}{\varepsilon}\exp\left[\frac{d-x}{\sqrt{2}\varepsilon}\right]. \tag{2.14}$$

For the second derivative using (2.2c) and Peano formula one obtains
$$\left|\partial_{xx}\hat{h}_\varepsilon(x, P)\right| = \left|\Pi_\varepsilon(\hat{h}_\varepsilon(x, P)) - P\right| \leq \left|\Pi'_\varepsilon(\theta_\varepsilon(P))(\hat{h}_\varepsilon(x) - \hat{h}_\varepsilon^-(P))\right|,$$
where $\theta_\varepsilon(P)$ is a point in interval $\left(\hat{h}_\varepsilon^-, \hat{h}_\varepsilon(x, P)\right)$. Therefore, using again (2.12) one arrives at
$$|\partial_{xx}\hat{h}_\varepsilon(x, P)| \leq \frac{C_0}{\varepsilon^2}\exp\left[\frac{d-x}{\sqrt{2}\varepsilon}\right].$$

Analogously, one can derive estimates for $|\partial_x^k \hat{h}_\varepsilon(x, P)|$ with $k = 3, 4$. This together with (2.13)–(2.14) implies (2.8a)–(2.8b) in the case $x > d$.

Next, integrating the first integral (2.6) on a interval (η, x) with $0 < \eta < x$ one obtains
$$x - \eta = \int_{\hat{h}_\varepsilon(x, P)}^{\hat{h}_\varepsilon(\eta, P)} \frac{dh}{\sqrt{-2\mathcal{U}_\varepsilon(h, P)}}.$$

Differentiation of the last expression with respect to P, using of (2.6) and subsequent taking $\eta = x_\varepsilon^c(P)$, where a point $x_\varepsilon^c(P)$ is defined by
$$\hat{h}_\varepsilon(x_\varepsilon^c(P), P) := \hat{h}_\varepsilon^c(P),$$
yields
$$\partial_P \hat{h}_\varepsilon(x, P) = \frac{\partial_P \hat{h}_\varepsilon(x_\varepsilon^c(P), P)}{\partial_x \hat{h}_\varepsilon(x_\varepsilon^c(P), P)}\partial_x \hat{h}_\varepsilon(x, P) +$$
$$+ \partial_x \hat{h}_\varepsilon(x, P) \int_{\hat{h}_\varepsilon(x, P)}^{\hat{h}_\varepsilon^c(P)} \frac{(h - \hat{h}_\varepsilon^-(P))\, dh}{\sqrt{(-2\mathcal{U}_\varepsilon(h, P))^3}} \tag{2.15}$$

Using that $\mathcal{U}_\varepsilon(h, P)$ decreases for fixed ε, P on $(\hat{h}_\varepsilon^-(P), \hat{h}_\varepsilon^c(P))$ and again (2.12) one estimates
$$\left|\partial_x \hat{h}_\varepsilon(x, P) \int_{\hat{h}_\varepsilon(x, P)}^{\hat{h}_\varepsilon^c(P)} \frac{(h - \hat{h}_\varepsilon^-(P))\, dh}{\sqrt{(-2\mathcal{U}_\varepsilon(h, P))^3}}\right| = \int_{\hat{h}_\varepsilon(x, P)}^{\hat{h}_\varepsilon^c(P)} \frac{(h - \hat{h}_\varepsilon^-(P))}{-2\mathcal{U}_\varepsilon(h, P)}\sqrt{\frac{\mathcal{U}_\varepsilon(\hat{h}_\varepsilon(x, P), P)}{\mathcal{U}_\varepsilon(h, P)}}\, dh$$
$$\leq \int_{\hat{h}_\varepsilon(x, P)}^{\hat{h}_\varepsilon^c(P)} \frac{dh}{2\Pi'_\varepsilon(\theta_\varepsilon(P))\left(h - \hat{h}_\varepsilon^-(P)\right)} \leq \varepsilon^2 \ln\left(\frac{\hat{h}_\varepsilon^c(P) - \hat{h}_\varepsilon^-(P)}{\hat{h}_\varepsilon(x, P) - \hat{h}_\varepsilon^-(P)}\right)$$
$$\leq -\varepsilon^2 \ln\left(\hat{h}_\varepsilon(x, P) - \hat{h}_\varepsilon^-(P)\right) \leq C_2\, \varepsilon(x - d),$$

where constant C_2 does not depend on ε, P. In the last expression we also used asymptotics (2.3a), (2.5) and estimate (2.13). Next, using $\Pi_\varepsilon(\hat{h}_\varepsilon^c(P)) - P = 0$ one obtains

$$\left|\frac{\partial_P \hat{h}_\varepsilon(x_\varepsilon^c(P), P)}{\partial_x \hat{h}_\varepsilon(x_\varepsilon^c(P), P)}\right| \leq C_3$$

where constant C_3 does not depend on ε, P. Therefore, using (2.14) one obtains

$$\left|\frac{\partial_P \hat{h}_\varepsilon(x_\varepsilon^c(P), P)}{\partial_x \hat{h}_\varepsilon(x_\varepsilon^c(P), P)} \partial_x \hat{h}_\varepsilon(x, P)\right| \leq \frac{C_3 C_0}{\varepsilon} \exp\left[\frac{d-x}{\sqrt{2}\varepsilon}\right]$$

The last three estimate imply (2.8c) in the case $x > d$. The case $x < -d$ for (2.8a)–(2.8c) can be shown analogously using that $\hat{h}_\varepsilon(x, P)$ and $\partial_P \hat{h}_\varepsilon(x, P)$ are odd functions in x. ∎

2.1.2 Stationary Solutions for the Strong-slip and Free Films Models

Here we derive a new result on stationary solutions of model (1.3a)–(1.3b), which turn out to be analogous to ones of Theorem 2.1 above.

Proposition 2.3. *System (1.3a)–(1.3b) considered on the whole real line $\mathbb{R}$ has a family of steady states parameterized by a parameter $P \in (0, P_{max})$, where P_{max} is defined in (2.1), with positive nonconstant height profile given by $\hat{h}_\varepsilon(x, P)$ and identically zero velocity.*

Proof: Steady states to (1.3a)–(1.3b) with a positive height profile are described by

$$\mathrm{Re}\, hu\, \partial_x u = 4\partial_x(h\partial_x u) + h\, \partial_x(\partial_{xx} h - \Pi_\varepsilon(h)) - \frac{u}{\beta},$$

$$0 = -\partial_x(hu).$$

By direct substitution and using (2.2a) one can check that $[\hat{h}_\varepsilon(x, P), 0]$ with $P \in (0, P_{max})$ form a family of stationary solutions to (1.3a)–(1.3b) on $\mathbb{R}$. ∎

Remark 2.4. Following the lines of the proof for Proposition 2.3 one can easily see that all the assertions of it hold for the suspended free films model (1.7a)–(1.7b) as well, what is natural, because as it was stated in Chapter 1 the later one is a limiting case for (1.3a)–(1.3b) as slip length $\beta \to 0$. Finally, Theorem 2.1 and Proposition 2.3 together state that all lubrication models considered on $\mathbb{R}$ possess similar families of positive nonconstant stationary solutions. ∎

2.2 Near-equilibrium Solutions and Generalized Gradient Flow

It is well-known that the driving forces that underly the initial dewetting scenario of a thin film, from rupture towards formation of complex fluid patterns, are intermolecular forces. This has been shown in the framework of the no-slip or weak-slip lubrication models, see e.g. Williams and Davis [13]. In fact other lubrication models show similar phases of the initial dewetting scenario, where now interfacial slip has an important influence on the morphology of the resulting patterns and the time scale on which they evolve, see Münch et al. [1], Peschka [28], Peschka et al. [43] for detailed analysis. However, as has been discussed by Glasner and Witelski [2] intermolecular forces are also important in the late phases when arrays of near-equilibrium droplets have formed, connected by a thin layer whose height is determined by competition between van-der-Waals attractive and Born repulsive forces. Here, the small flux across this layer plays an important

Chapter 2 Asymptotical Derivation of Reduced ODE Models

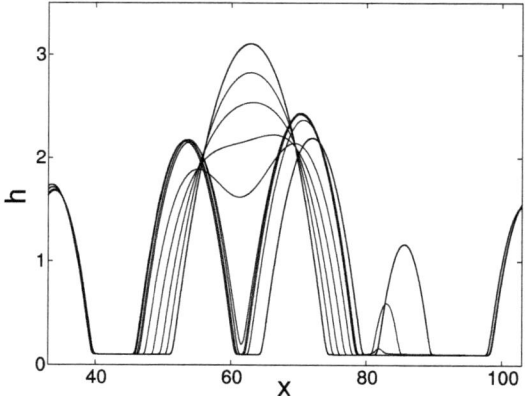

Figure 2.2: Numerical solution to (1.3a)–(1.3b) with $\varepsilon = 0.1$, $\beta = 2.5$ showing an example of a coarsening process (collapse of the 4th small droplet and collision of 2nd and 3rd ones) in the array of five quasiequilibrium droplets.

role in the coarsening dynamics of these arrays of droplets, where the central part of each droplet is nearly an equilibrium solution we have just discussed in the previous paragraphs. When Re is sufficiently small, two components of the coarsening regime can be identified, see Glasner et al. [3], Glasner and Witelski [29], namely collapse and collision (see example in Figure 2.2). One can qualitatively explain the driving effects for collapse and collision using presence of a generalized gradient flow structure. As it is found by Bertozzi et al. [15] the functional

$$E(h) = \int_{-L}^{L} U_\varepsilon(h) + \frac{(\partial_x h)^2}{2} \, dx \qquad (2.16)$$

is a Lyapunov functional for (1.6) with the boundary conditions (1.10), where $U_\varepsilon(h)$ is given by (1.5). Following the proof in Bertozzi et al. [15] one can easily generalize the result, i.e. (2.16) is a Lyapunov functional for the general mobility model (1.9) with the boundary conditions (1.10). Analogously, we find here a Lyapunov functional for the strong-slip system (1.3a)–(1.3b) and its limiting case (1.7a)–(1.7b), namely we prove the following proposition.

Proposition 2.5. *A functional*

$$E(u, h) = \int_{-L}^{L} U_\varepsilon(h) + \frac{Re}{2} h u^2 + \frac{(\partial_x h)^2}{2} \, dx \qquad (2.17)$$

is a Lyapunov functional for the system (1.3a)–(1.3b) (and for (1.7a)–(1.7b) as well) with boundary conditions (1.12)–(1.13).

Proof: To prove that $E(h, u)$ is a Lyapunov functional we show that for any solution $[h(x, t), u(x, t)]$ of (1.3a)–(1.3b) with (1.12)–(1.13) one has $dE(h(x,t), u(x,t))/dt \leq 0$.
We note that from integration by parts and using (1.13) one obtains

$$\frac{dE}{dt} = \text{Re} \int_{-L}^{L} u h \, \partial_t u \, dx + \int_{-L}^{L} (\Pi_\varepsilon(h) - \partial_{xx} h) \, \partial_t h \, dx + \text{Re} \int_{-L}^{L} \frac{u^2}{2} \partial_t h \, dx \,.$$

Substitution of $\partial_t h$ from (1.3b), integration by parts of the second and third term and noting the velocity boundary conditions result in

$$\frac{dE}{dt} = \text{Re}\int_{-L}^{L} uh\, \partial_t u\, dx + \int_{-L}^{L} hu\, \partial_x \left(\Pi_\varepsilon(h) - \partial_{xx}h\right) dx + \text{Re}\int_{-L}^{L} (hu)\, u\, \partial_x u\, dx\,. \tag{2.18}$$

Recall that from equation (1.3a)

$$-\partial_x \left(\partial_{xx}h - \Pi_\varepsilon(h)\right) = -\text{Re}\left(\partial_t u + u\partial_x u\right) + \frac{4}{h}\partial_x(h\partial_x u) - \frac{u}{\beta h}\,.$$

Using this in (2.18) we obtain

$$\frac{dE}{dt} = \int_{-L}^{L}\left(4\partial_x(h\partial_x u)\,u - \frac{u^2}{\beta}\right) dx\,.$$

Integration by parts of the first term and noting (1.12) gives

$$\frac{dE}{dt} = -\int_{-L}^{L} 4h\,(\partial_x u)^2\,dx - \int_{-L}^{L}\frac{u^2}{\beta}\,dx \leq 0 \text{ for all } t > 0, \tag{2.19}$$

provided $h(x,t) > 0$. ∎

Lyapunov functionals (2.16) and (2.17) induce a generalized gradient flow structure for (1.8) and (1.3a)–(1.3b), respectively. During the coarsening process in an array of droplets driven by (1.3a)–(1.3b) the energy (2.17) dissipates with the rate given by (2.19). As discussed in detail in Glasner and Witelski [2] and Otto et al. [37] in such a framework the collapse component of the coarsening process is driven by the dissipation of energy because the energy of two droplets before collapse is greater than one of the remaining after it droplet and its surrounding thin layer. In turn the migration of droplets can be explained by the presence of the Raleigh principle associated with the generalized gradient structure. It implies the non-zero mass flux between droplets in an array and in fact gives a rise for the motion of droplets and consequently for the collision component of the coarsening process. The same description of the two coarsening effects hold for system governed by the general mobility model (1.9), but now with (2.16).

2.3 Asymptotical Derivation of Reduced Model for One Droplet

2.3.1 Derivation for the General Mobility Model

The description of the slow motion of near-equilibrium droplets in terms of the equilibrium solutions, like those we have discussed in section 2.1, with parameters characterizing the position of the center of the droplet and its pressure that vary on that slow time scale, was first shown by Glasner and Witelski [2] for the no-slip lubrication equation. Based on their work we generalize and briefly describe the corresponding asymptotical approach for the general mobility model and cover in this way, besides no-slip, also weak and intermediate-slip cases.

The flux in the case of the general mobility model is defined as as

$$J = M(h)\partial_x\left(\partial_{xx}h - \Pi_\varepsilon(h)\right).$$

We consider the evolution of one droplet governed by the model (1.8) on the interval $[-\tilde{L}, \tilde{L}]$ at the boundary of which fluxes

$$J(-\tilde{L},t) = J_-(t),\ J(+\tilde{L},t) = J_+(t),$$

Chapter 2 Asymptotical Derivation of Reduced ODE Models

are imposed, where $J_\pm(t) = \sigma \bar{J}_\pm(t)$ and $\sigma \ll 1$. The fourth-order boundary value problem is complete by requiring two more boundary conditions

$$\partial_x h(-\tilde{L}, t) = 0, \quad \partial_x h(\tilde{L}, t) = 0. \tag{2.20}$$

Since the flux is very small, i.e. of order $\sigma \ll 1$, one can assume the solution $h(\cdot, t)$ (at every fixed time t) to have the form of a perturbed stationary solution from Theorem 2.1, initially centered at $x = \xi_0$ with initial pressure P_0, i.e $h(x,0) \approx \hat{h}_\varepsilon(x - \xi_0, P_0)$ restricted to the interval $(-\tilde{L}, \tilde{L})$, and evolving on a slow-time scale

$$\tau = \sigma t. \tag{2.21}$$

Following Glasner and Witelski [2] we make an asymptotical ansatz

$$h(x,\tau) = \hat{h}_\varepsilon(x - \xi(\tau), P(\tau)) + \sigma h_1(x,\tau) + O(\sigma^2), \tag{2.22}$$

where the position $\xi(\tau)$ and the pressure $P(\tau)$ of the droplet vary slowly in time. Below everywhere in this and the next section we denote

$$\hat{h}_\varepsilon := \hat{h}_\varepsilon(x - \xi, P), \hat{h}_\varepsilon^- := \hat{h}_\varepsilon^-(P), \hat{h}_\varepsilon^+ := \hat{h}_\varepsilon^-(P). \tag{2.23}$$

Substituting (2.22) in (1.9) we obtain to the leading order in σ

$$-\partial_x \hat{h}_\varepsilon \frac{d\xi}{d\tau} + \partial_P \hat{h}_\varepsilon \frac{dP}{d\tau} = \mathcal{L} h_1, \tag{2.24}$$

where $\mathcal{L}$ is a linear differential operator given by

$$\mathcal{L} h := \partial_x \left(M(\hat{h}_\varepsilon) \partial_x \left[\Pi_\varepsilon'(\hat{h}_\varepsilon) h - \partial_{xx} h \right] \right),$$

and the perturbation h_1 satisfies boundary conditions (2.20) and flux conditions

$$-M(\hat{h}_\varepsilon) \left(\partial_x \left[\Pi_\varepsilon'(\hat{h}_\varepsilon) h_1(-\tilde{L}) - \partial_{xx} h_1(-\tilde{L}) \right] \right) = \bar{J}_-,$$
$$-M(\hat{h}_\varepsilon) \left(\partial_x \left[\Pi_\varepsilon'(\hat{h}_\varepsilon) h_1(\tilde{L}) - \partial_{xx} h_1(\tilde{L}) \right] \right) = \bar{J}_+. \tag{2.25}$$

We introduce then formally an adjoint operator $\mathcal{L}^*$,

$$\mathcal{L}^* g := \left(\Pi_\varepsilon'(\hat{h}_\varepsilon) - \partial_{xx} \right) \left[\partial_x(M(\hat{h}_\varepsilon) \partial_x g) \right],$$

the kernel of which is spanned by two functions

$$g_1(x) := 1, \quad \text{and} \quad g_2(x) := \int_0^x \frac{\hat{h}_\varepsilon - \hat{h}_\varepsilon^-}{M(\hat{h}_\varepsilon)} \, dx'.$$

Using these functions and a suitable change of variables so that the linear problem with non-homogeneous boundary conditions (2.25) is transformed to one with homogeneous boundary conditions, one can impose two necessary conditions (similar to that used in Fredholm alternative) on the solvability of (2.24), which result in the system of ODEs for the pressure and position of the droplet, written in the original time scale as

$$\frac{dP}{dt} = C_P(J_+ - J_-), \quad \frac{d\xi}{dt} = -C_\xi(J_+ + J_-), \tag{2.26}$$

18

where
$$C_P = \frac{1}{-\int_{-\tilde{L}}^{\tilde{L}} \partial_P \hat{h}_\varepsilon \, dx}. \tag{2.27}$$

The only difference to the no-slip case described in Glasner and Witelski [2] and the dependence on the mobility $M(h)$ is found in the motion coefficient

$$C_\xi = \frac{\int_{-\tilde{L}}^{\tilde{L}} \dfrac{\hat{h}_\varepsilon - \hat{h}_\varepsilon^-}{M(\hat{h}_\varepsilon)} \, dx}{\int_{-\tilde{L}}^{\tilde{L}} \dfrac{\left(\hat{h}_\varepsilon - \hat{h}_\varepsilon^-\right)^2}{M(\hat{h}_\varepsilon)} \, dx}. \tag{2.28}$$

In the next section we propose algorithm for the integration of (2.27) and (2.28), along with coefficients for the dimension-reduced model for the strong-slip case given in the next paragraph.

2.3.2 Derivation for the Strong-slip and Free Suspended Films Models

Here we modify the asymptotical approach from the previous paragraph to make it applicable to two equation systems like (1.3a)–(1.3b) and its limiting case (1.7a)–(1.7b). Let us start from the strong-slip model. We again describe the evolution of one droplet on an interval $[-\tilde{L}, \tilde{L}]$ governed now by the system (1.3a)–(1.3b). We restrict our derivation to the regime where

$$\sigma^2 Re \ll 1, \tag{2.29}$$

i.e. for sufficiently small Re numbers in (1.3a)–(1.3b), and assume that again a single droplet behaves on a slow time scale τ given by (2.21) and can be parameterized by slow evolution of its pressure $P(\tau)$ and position $\xi(\tau)$. We make the following asymptotic ansatz:

$$\begin{aligned} h(x,\tau) &= \hat{h}_\varepsilon(x - \xi(\tau), P(\tau)) + \sigma h_1(x,\tau) + O(\sigma^2), \\ u(x,\tau) &= \sigma u_1(x,\tau) + O(\sigma^2). \end{aligned} \tag{2.30}$$

At the boundary of the interval we impose flux conditions, which can be written in terms of velocity at the boundary:

$$u(\pm \tilde{L}, \tau) = \sigma u_\pm(\tau). \tag{2.31}$$

In this case we define the fluxes $J_\pm$ imposed on the droplet as,

$$J_\pm := \sigma \hat{h}_\varepsilon^- u_\pm. \tag{2.32}$$

Below we use again notation (2.23). After substitution of (2.22) into (1.3a)–(1.3b) and noting (2.29) to the first order in σ we obtain (in matrix notation)

$$\begin{bmatrix} 0 \\ -\partial_x \hat{h}_\varepsilon \dfrac{d\xi}{d\tau} + \partial_P \hat{h}_\varepsilon \dfrac{dP}{d\tau} \end{bmatrix} = \mathcal{L} \begin{bmatrix} h_1 \\ u_1 \end{bmatrix}, \tag{2.33}$$

where $\mathcal{L}$ is a linear differential operator given by

$$\mathcal{L} \begin{bmatrix} h \\ u \end{bmatrix} = \begin{bmatrix} 4\, \partial_x(\hat{h}_\varepsilon \partial_x u) + \hat{h}_\varepsilon \partial_x(\partial_{xx} h - h \Pi'_\varepsilon(\hat{h}_\varepsilon)) - \dfrac{u}{\beta} \\ -\partial_x(\hat{h}_\varepsilon u) \end{bmatrix}.$$

Chapter 2 Asymptotical Derivation of Reduced ODE Models

The velocity correction term $u_1(x, \tau)$ satisfies according to (2.31)

$$u_1(\pm \tilde{L}, \tau) = u_{\pm}(\tau). \tag{2.34}$$

Formally, the adjoint operator to $\mathcal{L}$ is

$$\mathcal{L}^* \begin{bmatrix} g \\ v \end{bmatrix} = \begin{bmatrix} 4\partial_x(\hat{h}_\varepsilon \partial_x g) - \dfrac{g}{\beta} + \hat{h}_\varepsilon \partial_x v \\ \left(\Pi'_\varepsilon(\hat{h}_\varepsilon) - \partial_{xx} \right) \partial_x(\hat{h}_\varepsilon g) \end{bmatrix}.$$

It is easy to see that the kernel of it is spanned by two functions,

$$\begin{bmatrix} g_1 \\ v_1 \end{bmatrix} := \begin{bmatrix} 0 \\ 1 \end{bmatrix} \tag{2.35}$$

and

$$\begin{bmatrix} g_2 \\ v_2 \end{bmatrix} := \begin{bmatrix} \dfrac{\hat{h}_\varepsilon - \hat{h}_\varepsilon^-}{\hat{h}_\varepsilon} \\ \displaystyle\int_0^x \dfrac{\hat{h}_\varepsilon - \hat{h}_\varepsilon^-}{\beta \hat{h}_\varepsilon^2} - T(\hat{h}_\varepsilon, \partial_{x'}\hat{h}_\varepsilon, \partial_{x'x'}\hat{h}_\varepsilon)\, dx' \end{bmatrix}, \tag{2.36}$$

where

$$T(\hat{h}_\varepsilon, \partial_x \hat{h}_\varepsilon, \partial_{xx}\hat{h}_\varepsilon) := 4\, \hat{h}_\varepsilon^- \left(\dfrac{\hat{h}_\varepsilon \partial_{xx}\hat{h}_\varepsilon - (\partial_x \hat{h}_\varepsilon)^2}{\hat{h}_\varepsilon^3} \right). \tag{2.37}$$

Again using (2.35)–(2.36) and a transformation of the linear problem (2.33) with nonhomogeneous boundary conditions (2.34) to a problem with homogeneous boundary conditions, one can impose two necessary conditions on the solvability of (2.24) which result in ODEs (2.40a)–(2.40b). Our derivation below is equivalent to this procedure. In transformations below by Proposition 2.2 several boundary terms arising after subsequent applications of integration by parts are transcendentally small as $\varepsilon \to 0$. Therefore, we skip them and write an equivalence sign $\sim$ (see Definition 1.1) instead of $=$.

To derive an equation for $P(\tau)$ we multiply the second line in the matrix equation (2.33) by v_1 and integrate over the interval $[-\tilde{L}, \tilde{L}]$. We can make use of the fact that $\partial_x \hat{h}_\varepsilon$ is an odd function in x and $\int_{-\tilde{L}}^{\tilde{L}} \partial_x(u_1 \hat{h}_\varepsilon)\, dx \sim \hat{h}_\varepsilon^-(u_+ - u_-)$. This leads us to the equation

$$\dfrac{dP}{d\tau} \sim -\left(\int_{-\tilde{L}}^{\tilde{L}} \partial_P \hat{h}_\varepsilon\, dx \right)^{-1} \hat{h}_\varepsilon^-(u_+ - u_-),$$

the leading order of which, written in the original time scale and using (2.32), gives (2.40a) with the coefficient (2.41).

Similarly, we multiply the second line in the matrix equation (2.33) by v_2 and integrate over the interval $[-\tilde{L}, \tilde{L}]$. Again using odd or even symmetry we get

$$-\dfrac{d\xi}{d\tau} \int_{-\tilde{L}}^{\tilde{L}} v_2\, \partial_x \hat{h}_\varepsilon\, dx = -\int_{-\tilde{L}}^{\tilde{L}} v_2\, \partial_x(\hat{h}_\varepsilon u_1)\, dx,$$

which transforms to

$$\dfrac{d\xi}{d\tau} = \dfrac{\int_{-\tilde{L}}^{\tilde{L}} v_2\, \partial_x(\hat{h}_\varepsilon u_1)\, dx}{\int_{-\tilde{L}}^{\tilde{L}} v_2\, \partial_x \hat{h}_\varepsilon\, dx}. \tag{2.38}$$

Next, we calculate integrals in the denominator and numerator in the last expression. Denote

the numerator as
$$I_1 = \int_{-\tilde{L}}^{\tilde{L}} v_2 \, \partial_x(\hat{h}_\varepsilon u_1) \, dx.$$

Using definition (2.37), two times integration by parts and the fact that $v_2(x)$ is an odd function, one obtains

$$I_1 \sim \hat{h}_\varepsilon^- v_2(\tilde{L})(u_+ + u_-) - \int_{-\tilde{L}}^{\tilde{L}} u_1 \left(\frac{\hat{h}_\varepsilon - \hat{h}_\varepsilon^-}{\beta \hat{h}_\varepsilon} - \hat{h}_\varepsilon T(\hat{h}_\varepsilon, \partial_x \hat{h}_\varepsilon, \partial_{xx} \hat{h}_\varepsilon) \right) dx$$

$$\sim \hat{h}_\varepsilon^- v_2(\tilde{L})(u_+ + u_-) - \int_{-\tilde{L}}^{\tilde{L}} u_1 \left(\frac{\hat{h}_\varepsilon - \hat{h}_\varepsilon^-}{\beta \hat{h}_\varepsilon} \right) dx - 4 \int_{-\tilde{L}}^{\tilde{L}} \frac{\partial_x u_1 \, \partial_x \hat{h}_\varepsilon \, \hat{h}_\varepsilon^-}{\hat{h}_\varepsilon} \, dx.$$

Once more integration by parts and making use of the first line from (2.33) gives us

$$I_1 \sim \hat{h}_\varepsilon^- v_2(\tilde{L})(u_+ + u_-) - \int_{-\tilde{L}}^{\tilde{L}} u_1 \left(\frac{\hat{h}_\varepsilon - \hat{h}_\varepsilon^-}{\beta \hat{h}_\varepsilon} \right) dx - \left[4 \partial_x u_1 (\hat{h}_\varepsilon - \hat{h}_\varepsilon^-) \right]_{x=\pm \tilde{L}}$$

$$+ \int_{-\tilde{L}}^{\tilde{L}} \frac{\hat{h}_\varepsilon - \hat{h}_\varepsilon^-}{\hat{h}_\varepsilon} 4 \partial_x(\hat{h}_\varepsilon u_1) \, dx$$

$$\sim \hat{h}_\varepsilon^- v_2(\tilde{L})(u_+ + u_-) + \int_{-\tilde{L}}^{\tilde{L}} \frac{\hat{h}_\varepsilon - \hat{h}_\varepsilon^-}{\hat{h}_\varepsilon} \left(4 \partial_x(\hat{h}_\varepsilon \partial_x u_1) - \frac{u_1}{\beta} \right) dx$$

$$\sim \hat{h}_\varepsilon^- v_2(\tilde{L})(u_+ + u_-) - \int_{-\tilde{L}}^{\tilde{L}} (\hat{h}_\varepsilon - \hat{h}_\varepsilon^-) \, \partial_x \left(\partial_{xx} h_1 - \Pi'_\varepsilon(\hat{h}_\varepsilon) h_1 \right) dx.$$

Integrating further by parts three times one obtains finally

$$I_1 \sim \hat{h}_\varepsilon^- v_2(\tilde{L})(u_+ + u_-) + \int_{-\tilde{L}}^{\tilde{L}} \partial_x \hat{h}_\varepsilon \left(\partial_{xx} h_1 - \Pi'_\varepsilon(\hat{h}_\varepsilon) h_1 \right) dx$$

$$\sim \hat{h}_\varepsilon^- v_2(\tilde{L})(u_+ + u_-) + \int_{-\tilde{L}}^{\tilde{L}} h_1 \left(\partial_{xx} - \Pi'_\varepsilon(\hat{h}_\varepsilon) \right) \partial_x \hat{h}_\varepsilon \, dx$$

$$\sim \hat{h}_\varepsilon^- v_2(\tilde{L})(u_+ + u_-).$$

Hence, using the odd symmetry of $v_2(x)$, we can write

$$\int_{-\tilde{L}}^{\tilde{L}} v_2 \, \partial_x(\hat{h}_\varepsilon u) \, dx = \frac{1}{2} \left(\int_{-\tilde{L}}^{\tilde{L}} \frac{\hat{h}_\varepsilon - \hat{h}_\varepsilon^-}{\beta \hat{h}_\varepsilon} - \hat{h}_\varepsilon T(\hat{h}_\varepsilon, \partial_x \hat{h}_\varepsilon, \partial_{xx} \hat{h}_\varepsilon) \, dx \right) \hat{h}_\varepsilon^- (u_+ + u_-). \tag{2.39}$$

The denominator of (2.38) can be written as

$$\int_{-\tilde{L}}^{\tilde{L}} v_2 \, \partial_x \hat{h}_\varepsilon \, dx \sim - \int_{-\tilde{L}}^{\tilde{L}} \hat{h}_\varepsilon \, \partial_x v_2 \, dx + \hat{h}_\varepsilon^- (v_2(\tilde{L}) - v_2(-\tilde{L})) \sim - \int_{-\tilde{L}}^{\tilde{L}} (\hat{h}_\varepsilon - \hat{h}_\varepsilon^-) \, \partial_x v_2 \, dx.$$

Substituting this and (2.39) with $v_2(x)$ given by (2.36) into (2.38) and taking the leading order of the latter one as $\varepsilon \to 0$ results in equation (2.40b) with (2.42). We summarize that two evolution equations for droplet's pressure $P(t)$ and position $\xi(t)$, written in original time scale, finally take form

Chapter 2 Asymptotical Derivation of Reduced ODE Models

$$\frac{dP}{dt} = C_P \left(J_+ - J_- \right), \tag{2.40a}$$

$$\frac{d\xi}{dt} = - C_\xi \left(J_+ + J_- \right), \tag{2.40b}$$

where

$$C_P := \frac{1}{-\int_{-\tilde{L}}^{\tilde{L}} \partial_P \hat{h}_\varepsilon \, dx} \tag{2.41}$$

and a so called *mobility coefficient*

$$C_\xi := \frac{\int_{-\tilde{L}}^{\tilde{L}} \left(\frac{\hat{h}_\varepsilon - \hat{h}_\varepsilon^-}{\hat{h}_\varepsilon^2} - \beta T(\hat{h}_\varepsilon, \partial_x \hat{h}_\varepsilon, \partial_{xx} \hat{h}_\varepsilon) \right) dx}{2 \int_{-\tilde{L}}^{\tilde{L}} \left(\frac{(\hat{h}_\varepsilon - \hat{h}_\varepsilon^-)^2}{\hat{h}_\varepsilon^2} - \beta T(\hat{h}_\varepsilon, \partial_x \hat{h}_\varepsilon, \partial_{xx} \hat{h}_\varepsilon)(\hat{h}_\varepsilon - \hat{h}_\varepsilon^-) \right) dx}. \tag{2.42}$$

Coefficients (2.41)–(2.42) depend on droplet pressure P and position ξ through the given stationary height profile $\hat{h}_\varepsilon(x - \xi, P)$, its minimum $\hat{h}_\varepsilon^-(P)$ and function $T(\hat{h}_\varepsilon, \partial_x \hat{h}_\varepsilon, \partial_{xx} \hat{h}_\varepsilon)$ defined in (2.37). We observe that (2.42) differs from the corresponding coefficient (2.28) for the general mobility model and depends now on the slip-length β. In the next section we introduce an algorithm for integration of the integrals in (2.42). Moreover, we derive asymptotics for them in the limit $\varepsilon \to 0$, which implies that the denominator in of (2.42) is positive for all slip-lengths $\beta \geq 0$, but the numerator can change its sign with β.

Remark 2.6. For the case of negligible Trouton viscosity (when $\beta \to 0$) the function $\beta T(\hat{h}_\varepsilon, \partial_x \hat{h}_\varepsilon, \partial_{xx} \hat{h}_\varepsilon) \to 0$ and we recover reduced ODE model (2.26)–(2.28) for the intermediate slip case. On the other hand when $\beta \to \infty$, we obtain a reduced ODE system for lubrication model (1.7a)–(1.7b) describing free suspended films for which the coefficient (2.42) is replaced by

$$C_\xi = \frac{\int_{-\tilde{L}}^{\tilde{L}} T(\hat{h}_\varepsilon, \partial_x \hat{h}_\varepsilon, \partial_{xx} \hat{h}_\varepsilon) \, dx}{2 \int_{-\tilde{L}}^{\tilde{L}} T(\hat{h}_\varepsilon, \partial_x \hat{h}_\varepsilon, \partial_{xx} \hat{h}_\varepsilon)(\hat{h}_\varepsilon - \hat{h}_\varepsilon^-) \, dx}.$$

∎

We conclude that we derived in this section reduced finite-dimensional models describing evolution of one droplet for the complete set of lubrication equations stated in section 1.2. Whereas the pressure coefficient C_P is the same in all reduced models, the difference between them lies in the mobility coefficient C_ξ. Moreover, in contrast to slip cases described by general mobility model (1.9), in the strong-slip case the mobility coefficient (2.42) depends on the slip-length β. The consequences of this fact we investigate in Chapter 3.

2.4 Integration and Asymptotics for Coefficients C_P and C_ξ

First we suggest algorithm for a numerical integration of coefficients (2.27), (2.28), (2.42). Below K_i, $i = 1, ..., 12$ denote constants. We use notation (2.23) also here.

The contribution to the coefficient C_P to the leading order comes from the droplet core and we can neglect the contribution from the ultrathin film when $\epsilon \ll 1$. Therefore, it can be calculated

2.4 Integration and Asymptotics for Coefficients C_P and C_ξ

as was suggested first by Glasner and Witelski [2]:

$$C_P = \frac{1}{-\int_{-\tilde{L}}^{\tilde{L}} \partial_P \hat{h}_\varepsilon \, dx} \sim -\frac{3P^3}{4A^3}, \tag{2.43}$$

where

$$A := \sqrt{2|U_\varepsilon(\epsilon)|} = 1/\sqrt{3} \tag{2.44}$$

defines a so called droplet contact angle.

In contrast to this main contributions to coefficients C_ξ given by (2.28) or (2.42) comes from the ultra-thin layer. Below we describe integration of (2.42). In the same way coefficient (2.28) can be integrated. Using estimate (2.8a) from Proposition 2.2 we extend intervals in definition of (2.42) from $[-\tilde{L}, \tilde{L}]$ to $[-\infty, \infty]$. After this (2.42) is given by ratio of two improper integrals. In order to show that they converge we need to prove that integrals

$$I_{1,n} = \int_{-\infty}^{\infty} \frac{\left(\hat{h}_\varepsilon - \hat{h}_\varepsilon^-\right)^n}{\hat{h}_\varepsilon^2} \, dx, \quad I_{2,m} = \int_{-\infty}^{\infty} T(\hat{h}_\varepsilon, \partial_x \hat{h}_\varepsilon, \partial_{xx} \hat{h}_\varepsilon)(\hat{h}_\varepsilon - \hat{h}_\varepsilon^-)^m \, dx \tag{2.45}$$

converge for $n = 1, 2$ and $m = 0, 1$.

Let us start with $I_{1,n}$. Using the first integral (2.6) with (2.7) for the stationary solution $\hat{h}_\varepsilon$ and (2.2b) one can change variables in (2.45) and integrate both integrals over interval $\hat{h}_\varepsilon^- \leq h \leq \hat{h}_\varepsilon^+$:

$$I_{1,n} = 2 \int_{\hat{h}_\varepsilon^-}^{\hat{h}_\varepsilon^+} \frac{\left(h - \hat{h}_\varepsilon^-\right)^n}{h^2 \sqrt{-2\mathcal{U}_\varepsilon(h, P)}} \, dh.$$

These integrals are improper at both ends of the integration interval because $\mathcal{U}_\varepsilon(\hat{h}_\varepsilon^-, P) = \mathcal{U}_\varepsilon(\hat{h}_\varepsilon^+, P) = 0$. One can see that $\mathcal{U}_\varepsilon(h, P)$ is of a form:

$$\mathcal{U}_\varepsilon(h, P) = \frac{K_1 (h - \hat{h}_\varepsilon^-)^2 (h - \hat{h}_\varepsilon^+)(h - h_\varepsilon^n)}{h^3}, \tag{2.46}$$

where $h_\varepsilon^n < 0$ is the third negative zero of $\mathcal{U}_\varepsilon(h, P)$. Cancellation of $h - \hat{h}_\varepsilon^-$ in the denominator and the numerator makes $I_{1,n}$ improper only at the end $\hat{h}_\varepsilon^+$

$$I_{1,n} = K_2 \int_{\hat{h}_\varepsilon^-}^{\hat{h}_\varepsilon^+} \frac{\left(h - \hat{h}_\varepsilon^-\right)^{n-1}}{\sqrt{h(h - \hat{h}_\varepsilon^+)(h - h_\varepsilon^n)}} \, dh,$$

We assure that $I_{1,n}$ converges by making a second change of variables:

$$h = \hat{h}_\varepsilon^+ \cos(\theta). \tag{2.47}$$

Substituting this in $I_{1,n}$ yields

$$I_{1,n} = K_3 \int_0^{\mathrm{acos}(\hat{h}_\varepsilon^-/\hat{h}_\varepsilon^+)} \frac{\left(\hat{h}_\varepsilon^+ \cos(\theta) - \hat{h}_\varepsilon^-\right)^{n-1} \cos(\theta/2)}{\sqrt{\cos(\theta)(\cos(\theta) - h_\varepsilon^n/\hat{h}_\varepsilon^+)}} \, d\theta.$$

The last integral is proper and can be integrated by the three-point Gaussian quadrature.

Let us now calculate the second type of integrals, namely $I_{2,m}$ in (2.45). As before we use (2.6)

Chapter 2 Asymptotical Derivation of Reduced ODE Models

with (2.7) to change variables and integrate over $\hat{h}_\varepsilon^- \leq h \leq \hat{h}_\varepsilon^+$:

$$I_{2,m} = 2\int_{\hat{h}_\varepsilon^-}^{\hat{h}_\varepsilon^+} \frac{T(h, \partial_x h, \partial_{xx} h)(h - \hat{h}_\varepsilon^-)^m}{\sqrt{-2\mathcal{U}_\varepsilon(h,P)}} dh \quad \text{(using (2.37), (2.2a)-(2.2b) and (2.7))}$$

$$= 8\int_{\hat{h}_\varepsilon^-}^{\hat{h}_\varepsilon^+} \frac{\hat{h}_\varepsilon^-(h-\hat{h}_\varepsilon^-)^m(\Pi_\varepsilon(h)-P)}{h^2\sqrt{-2\mathcal{U}_\varepsilon(h,P)}} dh - 8\int_{\hat{h}_\varepsilon^-}^{\hat{h}_\varepsilon^+} \frac{\hat{h}_\varepsilon^-\left(h-\hat{h}_\varepsilon^-\right)^m\sqrt{-2\mathcal{U}_\varepsilon(h,P)}}{h^3} dh.$$

$$=: 8\hat{h}_\varepsilon^- \left(I_{2,m}^1 + I_{2,m}^2 \right) \tag{2.48}$$

The second integral $I_{2,m}^2$ in (2.48) is a proper integral at both ends of the integration interval and therfore converges. But the first integral $I_{2,m}^1$ is improper at the both ends. Note that the term $\Pi_\varepsilon(h) - P$ has the following form:

$$\Pi_\varepsilon(h) - P = \frac{K_4(h-\hat{h}_\varepsilon^-)(h-\hat{h}_\varepsilon^c)(h^2 + (h_\varepsilon^a)^2)}{h^4}$$

where $\hat{h}_\varepsilon^c$ has asymptotics (2.5) and h_ε^a is the modulus of two conjugate complex roots of algebraic equation $\Pi_\varepsilon(h) = P$. Using (2.46) and the last expression one can simplify $I_{2,m}^1$ as follows:

$$I_{2,m}^1 = K_5 \int_{\hat{h}_\varepsilon^-}^{\hat{h}_\varepsilon^+} \frac{(h-\hat{h}_\varepsilon^-)^m (h - \hat{h}_\varepsilon^c)(h^2+(h_\varepsilon^a)^2)}{\sqrt{h^9(h-\hat{h}_\varepsilon^+)(h-h_\varepsilon^n)}} dh.$$

Hence $I_{2,m}^1$ becomes proper at $\hat{h}_\varepsilon^-$. To make it proper at the right end $\hat{h}_\varepsilon^+$ we use again trigonometric change of variables (2.47) and proceed exactly as in case of $I_{1,n}$ above. This transforms $I_{2,m}^1$ to a proper integral that can be integrated by the three-point Gaussian quadrature. We have in summary that both integral in (2.45) converge, and therefore mobility coefficients (2.28) or (2.42) can be calculated numerically.

Besides the direct calculation of (2.45) one can estimate these integrals asymptotically in a limit $\epsilon \to 0$ following the similar approach in Appendix of Glasner and Witelski [29]. Such asymptotics we use in section 3.1 to estimate a critical slip-length β_{crit}. For these purposes we derive here the leading order asymptotics of integrals $I_{1,1}$ and $I_{2,m}$ for $m = 0, 1$ and prove that $I_{2,0} > 0$ and $I_{2,1} < 0$ for all sufficiently small ϵ.

Applying Taylor expansion to (2.7) in some neighborhoods $\mathcal{O}_\varepsilon^-$ and $\mathcal{O}_\varepsilon^+$ of $\hat{h}_\varepsilon^-$ and $\hat{h}_\varepsilon^+$, respectively, one obtains:

$$-\mathcal{U}_\varepsilon(h,P) \sim \begin{cases} \frac{1}{2}U_\varepsilon''(\hat{h}_\varepsilon^-)(h-\hat{h}_\varepsilon^-)^2 & \text{for all } h \in \mathcal{O}_\varepsilon^- \\ [P - U_\varepsilon'(\hat{h}_\varepsilon^+)](\hat{h}_\varepsilon^+ - h) & \text{for all } h \in \mathcal{O}_\varepsilon^+ \end{cases}. \tag{2.49}$$

Analogously, one can show:

$$\Pi_\varepsilon(h) - P \sim \Pi_\varepsilon(\hat{h}_\varepsilon^-)(h - \hat{h}_\varepsilon^-) \text{ for all } h \in \mathcal{O}_\varepsilon^-. \tag{2.50}$$

To estimate $I_{1,1}$, we note from (2.49) that the ratio $(h-\hat{h}_\varepsilon^-)/\sqrt{-2\mathcal{U}_\varepsilon(h,P)} \sim \text{const} > 0$ in $\mathcal{O}_\varepsilon^-$. Next, for $\hat{h}_\varepsilon^- \sim \varepsilon$, the factor $1/h^2$ makes the integrand relatively large in $\mathcal{O}_\varepsilon^-$. This contribution, along with the contribution from $\mathcal{O}_\varepsilon^+$ leads to an estimate

$$I_{1,1} \sim \frac{1}{\sqrt{U_\varepsilon''(\hat{h}_\varepsilon^-)}} \int_{\hat{h}_\varepsilon^-}^{\hat{h}_\varepsilon^c} \frac{dh}{h^2} + \frac{1}{\sqrt{2|P - U_\varepsilon'(\hat{h}_\varepsilon^+)|}} \int_{\hat{h}_\varepsilon^c}^{\hat{h}_\varepsilon^+} \frac{h - \hat{h}_\varepsilon^-}{h^2\sqrt{\hat{h}_\varepsilon^+ - h}} dh, \tag{2.51}$$

2.4 Integration and Asymptotics for Coefficients C_P and C_ξ

where the elliptic center point $\hat{h}_\varepsilon^c$ yields an effective cut-off for the influence of the behavior near $\hat{h}_\varepsilon^-$. Both integrals in (2.51) can be integrated analytically. Denoting them as $I_{1,1}^1$ and $I_{2,1}^2$, respectively, and using asymptotics (2.3a), (2.3b), (2.5) one obtains

$$I_{1,1}^1 = K_6 + O(\epsilon); \quad I_{1,1}^2 = K_7 \ln\left(\frac{2}{3\epsilon P}\right) + K_8 + O(\epsilon), \quad K_6, K_7 > 0.$$

Hence the final asymptotics for $I_{1,1}$ is of the form

$$I_{1,1} = K_7 \ln\left(\frac{2}{3\epsilon P}\right) + O(1). \tag{2.52}$$

Let us now estimate the integral $I_{2,0}$. To this end one needs to estimate integrals $I_{2,0}^1$ and $I_{2,0}^2$ in (2.48). Analogously to (2.51), using (2.49), (2.50) and definitions (1.4), (1.5) one writes

$$I_{2,0}^1 = \int_{\hat{h}_\varepsilon^-}^{\hat{h}_\varepsilon^+} \frac{\Pi_\varepsilon(h) - P}{h^2\sqrt{-2\mathcal{U}_\varepsilon(h,P)}} dh \sim \sqrt{\Pi_\varepsilon'(\hat{h}_\varepsilon^-)} \int_{\hat{h}_\varepsilon^-}^{\hat{h}_\varepsilon^c} \frac{dh}{h^2}$$
$$+ \frac{1}{\sqrt{2|P - U_\varepsilon'(\hat{h}_\varepsilon^+)|}} \int_{\hat{h}_\varepsilon^-}^{\hat{h}_\varepsilon^+} \frac{\Pi_\varepsilon(h) - P}{h^2\sqrt{\hat{h}_\varepsilon^+ - h}} dh,$$

where both integrals at the right-hand side can be integrated analytically. Again using asymptotics (2.3a), (2.3b), (2.5) one gets

$$I_{2,0}^1 = \frac{K_9}{\epsilon^2} + O(\epsilon), \quad K_9 > 0.$$

For the proper integral $I_{2,0}^2$ one obtains:

$$I_{2,0}^2 = \int_{\hat{h}_\varepsilon^-}^{\hat{h}_\varepsilon^+} \frac{\sqrt{-2\mathcal{U}_\varepsilon(h,P)}}{h^3} dh \leq \sqrt{-\mathcal{U}_\varepsilon(\hat{h}_\varepsilon^c, P)/2} \left(\frac{1}{\left(\hat{h}_\varepsilon^-\right)^2} - \frac{1}{\left(\hat{h}_\varepsilon^+\right)^2}\right)$$
$$= \frac{K_{10}}{\epsilon^2} + O\left(\frac{1}{\epsilon}\right) \quad K_{10} > 0.$$

Finally, last two asymptotics and (2.48) yield that integral $I_{2,0}$ is positive for sufficiently small ε and has the following asymptotics:

$$I_{2,0} = \frac{K_{11}}{\epsilon} + O(1), \quad K_{11} > 0. \tag{2.53}$$

One should notice that a similar estimation can be applied to integral $I_{2,1}$. The only difference here is that the integrand of $I_{2,1}^1$ is improper only at the end $h = \hat{h}_\varepsilon^+$. Hence the main contribution to $I_{2,1}^1$ comes from a neighborhood of $h = \hat{h}_\varepsilon^+$. The final asymptotics for $I_{2,1}$ is of the form:

$$I_{2,1} = K_{12} + O(\epsilon), \quad K_{12} < 0, \tag{2.54}$$

and hence $I_{2,1}$ is negative for sufficiently small $\varepsilon > 0$. We conclude that asymptotics (2.54) together with the fact that integrals $I_{1,n}$, $n = 0, 1$ defined in (2.45) are positive, because $\hat{h}_\varepsilon^- > \hat{h}_\varepsilon^- > 0$, imply that denominator of the mobility coefficient (2.42) is positive for all sufficiently small $\varepsilon > 0$.

Chapter 2 Asymptotical Derivation of Reduced ODE Models

2.5 Approximation for the Fluxes between Droplets

In this and next sections we consider evolution of an array of several droplets governed by one of lubrication models on a fixed interval $(-L, L)$ with boundary conditions (1.10) or (1.12)–(1.13) in the case of the general mobility model or the strong-slip model, respectively. By results of section 2.3 for each j-th droplet in such array correspond two ODEs which describe evolution of its pressure $P_j(t)$ and position $\xi_j(t)$ in time (see Figure 2.3). In order to couple equations for all

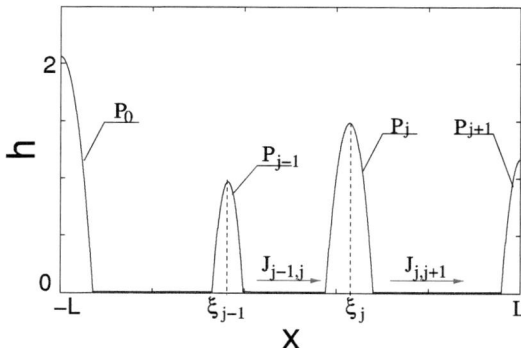

Figure 2.3: Geometric sketch for an array of several droplets

droplets in one ODE system and to describe coarsening process in droplet arrays we derive in this section asymptotic expressions for the fluxes $J_{j-1,j}$ and $J_{j,j+1}$ that j-th droplet experiences due to its neighbors. We start first with the intermediate-slip model (1.8). After deriving flux approximation in this case we state the result for no and weak slip models, which was already derived in Glasner and Witelski [2], but differs from one for the intermediate-slip model. Finally, we derive flux approximations for the strong-slip model (1.3a)–(1.3b).

2.5.1 Intermediate-slip Case

As for the no-slip case in Glasner and Witelski [2], we note first that the fluxes occur through the thin film of height $h = O(\varepsilon)$, connecting droplets, which are assumed to have a typical distance of $1/\delta$. Similarly, we obtain expression for the fluxes in the thin film by first scaling the variables to this *inner* region as follows

$$z = \delta x, \quad H = \frac{h}{\varepsilon}, \quad \tau = \sigma t,$$

where σ was introduced first in (2.21). For the intermediate-slip case one takes it as

$$\sigma = \varepsilon^2 \delta.$$

Substitution of this scaling in (1.8) yields

$$\varepsilon^2 \delta^{-1} H_\tau = \partial_z \left(H^2 \partial_z (U'(H) - \varepsilon^2 \delta^2 \partial_{zz} H) \right),$$

2.5 Approximation for the Fluxes between Droplets

with $U(H) := U_\varepsilon(h/\epsilon)$. Hence, to leading order as $\varepsilon \to 0$ the quasi stationary problem is

$$0 = \partial_{zz}[\mathcal{V}(H)],$$

where $\mathcal{V}(H)$ is defined by

$$\frac{d\mathcal{V}}{dH} = H^2 \frac{d^2 U}{dH^2} \quad \text{which is, in unscaled variables} \quad \frac{dV}{dh} = h^2 \frac{d^2 U_\varepsilon}{dh^2}. \tag{2.55}$$

One has $V(h) = \varepsilon \mathcal{V}(h/\varepsilon)$, so that the flux between droplets is $J = -\partial_x V(h)$.

Solving the *outer* boundary value problem $\partial_{xx} V(h) = 0$ with $V = V(\hat{h}_\varepsilon^-)$ at the apparent contact line, i.e. where the droplet merges into the ultrathin layer, we observe that as in Glasner and Witelski [2] for the no-slip case, also here we obtain that the flux between two neighboring droplets, labeled by j and $j+1$ is constant and given by their positions and pressures as

$$J_{j,j+1} = -\frac{V(\hat{h}_\varepsilon^-(P_{j+1})) - V(\hat{h}_\varepsilon^-(P_j))}{[\xi_{j+1} - A/P_{j+1}] - [\xi_j + A/P_j]}, \tag{2.56}$$

where A is defined in (2.44) and A/P_k is approximation for the droplet half-width (see Glasner and Witelski [2]). Note, that from (2.55) one has an explicite formula for function $V(h)$:

$$V = \frac{3\varepsilon^2}{h} - \frac{2\varepsilon^3}{h^2}. \tag{2.57}$$

Approximation in the no-slip and weak-slip cases can be derived in a similar way. The formula (2.56) is still valid in these cases, but in contrast to the intermediate-slip one the corresponding function $V(h)$ has a form:

$$V = -3\varepsilon^2 \log(h) - \frac{4\varepsilon^3}{h}.$$

2.5.2 Strong-slip Case

As we did throughout this chapter we neglect the inertial terms in the strong-slip model to obtain

$$0 = \frac{4}{h} \partial_x(h \partial_x u) + \partial_x (\partial_{xx} h - \Pi_\varepsilon(h)) - \frac{u}{\beta h},$$
$$\partial_t h = -\partial_x (hu).$$

In the derivation below we do not put any restrictions on the distance between droplets, because, in general, the reduced ODE system (2.40a)–(2.40b) should describe also dynamics of droplets, which not necessary locate far from each other (e.g. in order to approximate well migration towards collision). Therefore, we introduce only two scalings for the *inner* variables:

$$H = \frac{h}{\varepsilon}, \quad \tau = \sigma t.$$

These scalings then automatically imply the scaling for the velocity

$$W = \gamma u = \frac{\varepsilon}{\sigma} u,$$

27

because $u = J/h$ and flux scales like $1/\sigma$. The equations in *inner* scalings introduced above become

$$0 = \frac{4\sigma}{H}\partial_z(H\partial_z W) + \partial_z\left(\varepsilon^2 \partial_{zz}H - U'(H)\right) - \frac{\sigma}{\varepsilon\beta}\frac{W}{H}, \qquad (2.58a)$$

$$\varepsilon\partial_\tau H = -\partial_z(HW), \qquad (2.58b)$$

where again $U(H) := U_\varepsilon(h/\epsilon)$. From (2.58a) it is clear that surface tension term is negligible in comparison with other ones. Additionally, as for the no-slip and intermediate slip cases, we consider only the situation with constant flux, i.e. we neglect term $\varepsilon\,\partial_\tau H$ in the second equation. Therefore, the leading order equations for (2.58a)–(2.58b) are

$$0 = \frac{4\sigma}{H}\partial_z(H\partial_z W) - \partial_z(U'(H)) - \frac{\sigma}{\varepsilon\beta}\frac{W}{H}, \qquad 0 = \partial_z(HW),$$

Note that during derivation of the strong-slip model (1.3a)–(1.3b) in Münch et al. [1] the slip length β does not depend on $\varepsilon > 0$, so we need to consider only possible relations between scaling parameters σ and ε. Let us write

$$\sigma = \varepsilon^\lambda. \qquad (2.59)$$

Then the leading order system above transforms to

$$0 = \frac{4\varepsilon^\lambda}{H}\partial_z(H\partial_z W) - \partial_z(U'(H)) - \frac{\varepsilon^{\lambda-1}}{\beta}\frac{W}{H},$$
$$0 = \partial_z(HW),$$

Moreover, noting that in the *inner* region by asymptotics (2.3a) holding in the ultrathin layer between quasiequilibrium droplets, one has

$$H \sim 1 + \text{const}\,\varepsilon. \qquad (2.60)$$

Substituting this in $U(h)$ and applying Taylor expansion yield that $U'(H) \sim \text{const}\,\varepsilon$ as $\varepsilon \to 0$. Let us introduce

$$\widetilde{U}'(H) := U'(H)/\varepsilon = O(1).$$

The the system transforms to

$$0 = \frac{4\varepsilon^\lambda}{H}\partial_z(H\partial_z W) - \varepsilon\partial_z\left(\widetilde{U}'(H)\right) - \frac{\varepsilon^{\lambda-1}}{\beta}\frac{W}{H}, \qquad (2.61a)$$

$$0 = \partial_z(HW), \qquad (2.61b)$$

Looking at the last system one can distinguish three cases for balancing terms in (2.61a):
Case I($\lambda = 2$):

In this case Trouton term is negligible and the leading order system becomes

$$0 = \partial_z(\widetilde{U}'(H)) - \frac{W}{\beta H},$$
$$0 = \partial_z(HW),$$

This implies

$$HW = -\beta H^2 \partial_z\left(\widetilde{U}'(H)\right)$$

and
$$0 = \partial_z \left(H^2 \partial_z \left(U'(H) \right) \right).$$

If we define $\mathcal{V}(H)$ and $V(h)$ as in the intermediate-slip case and note that in *outer* scales
$$\frac{dV}{dh} \partial_x h = h^2 \frac{d^2 U_\varepsilon}{dh^2} \partial_x h, \tag{2.62}$$

then we can express the flux as
$$J = uh = -\beta h^2 \frac{d^2 U_\varepsilon}{dh^2} \partial_x h = -\beta \partial_x V. \tag{2.63}$$

Note, that by (2.62) we have the same expression (2.57) for $V(h)$ as in the intermediate case.
Case II($\lambda > 2$):
In this case both Trouton and the last term in (2.61a) are negligible in comparison with intermolecular potential. Therefore, the leading system becomes
$$0 = U''(H) \partial_z H,$$
$$0 = \partial_z (HW),$$

where $U''(H) = -3/H^4 + 4/H^5$. By (2.60) in the *inner* region $U''(H) \sim 1$ as $\varepsilon \to 0$. Therefore, solving the leading order system one obtains $H = \text{const}$ and $W = \text{const}$, but this contradicts to the matching with *outer* layer boundary conditions (see details below). Therefore, Case II is not possible.
Case II($\lambda < 2$):
In this case Trouton term and the intermolecular potential are negligible and we end up with a system
$$0 = \frac{W}{H},$$
$$0 = \partial_z (HW),$$

which leads again obviously to a contradiction.

We conclude that only Case I is possible. To give an explicit quasistationary formula for the fluxes between droplets it remains to get values of function (2.57) at the boundary of the contact line region. Here comes out *the main difference* between the strong-slip case and other slip regimes. Namely, in the formula (2.56) for other slip cases we matched directly the height profile h in the *inner* layer between droplets with its value $\hat{h}_\varepsilon^-$ given by (2.3a) at the border of the *outer* layer (i.e. parabolic droplet core) assuming that in passing the small contact line region function h and corresponding pressure profile to the leading order does not change. That means one just skips the contact line region in these cases. But as our numerical simulations show it is not allowed to do in the strong-slip case. Whereas in this case as in other slip-cases the pressure
$$p(x) := \Pi_\varepsilon(h(x)) - \partial_{xx} h(x)$$

is constant inside of droplet core (*outer* layer) and linear between droplets (*inner* layer) (see Figure 2.4), increasing slip-length from zero value one can observe additional pressure kinks in the contact line region between above layers. These pressure kinks obviously indicate changes in height profile $h(x)$ at the contact line region, and therefore the latter one should be incorporated in the matching process. If one denotes the contact line region corresponding to j-th droplet as $[a_j, b_j]$ (see Figure 2.4) then (2.63) holding in the *inner* layer $[b_j, b_{j+1}]$ and matching with the

Chapter 2 Asymptotical Derivation of Reduced ODE Models

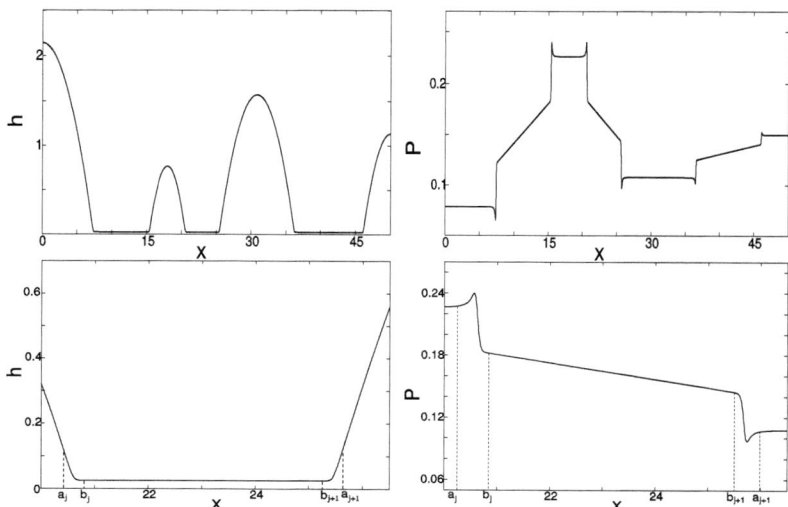

Figure 2.4: Upper row:Numerical example with $\varepsilon = 0.25$, $\beta = 10$, $L = 50$ of array of four droplets(left) and corresponding pressure profile(right). Lower row: Zoom of the interval between droplets(left) and corresponding pressure profile with kinks at contact lines(right)

boundary conditions at $x = b_j$ and $x = b_{j+1}$ results in the flux formula

$$J_{j,j+1} = -\beta \frac{V(h(b_{j+1})) - V(h(b_j))}{b_{j+1} - b_j}, \tag{2.64}$$

where $V(h)$ is given by (2.57). Although one knows, that at the end point of the *outer* layer height profile to the leading order is given by (2.3a), i.e. $h(a_j) \sim \hat{h}_\varepsilon^-(P_j)$, nevertheless the formula for $h(b_j)$ is not yet known, i.e. how height profile changes passing the contact line region $[a_j, b_j]$, and we can not use formula (2.64) so far for the flux approximation in the reduced ODE (2.40a)–(2.40b). The answer to this problems is out of the dissertation content and lies in appropriate scalings, asymptotical simplification and matching process for the strong-slip model in the contact line region, like those we did already in the *inner* layer. In the next sections for numerical simulations of the reduced ODE (2.40a)–(2.40b) we use approximation for fluxes

$$J_{j,j+1} = -\beta \frac{V(\hat{h}_\varepsilon^-(P_{j+1})) - V(\hat{h}_\varepsilon^-(P_j))}{[\xi_{j+1} - A/P_{j+1}] - [\xi_j + A/P_j]}. \tag{2.65}$$

as the closest one to the not fully determined strong-slip one (2.64). Comparison between reduced and lubrication models shows that when the slip length β increases the size of pressure kinks in the contact line region increases as well and flux approximation (2.65) works bad (see observation (iv) in paragraph 2.7.2). The fact that the reduced ODE works well for any value of slippage and the problem lies just in numerical approximations of fluxes between droplets is confirmed by observation (v) in paragraph 2.7.2.

2.6 Final Form of Reduced ODE Systems

We summarize here results of the previous three sections. Given an array of $N+1$ droplets (see Figure 2.3) on a bounded interval $[-L, L]$, coarsening process in which is governed by one of the lubrication equations defined in section 1.2 with boundary conditions (1.10) or (1.12)–(1.13), the evolution of pressures $P_j(t)$ and positions $\xi_j(t)$ for $j = 0, ..., N$ of droplets in the array is described by the following ODE system:

$$\frac{dP_j}{dt} = C_{P,j}(J_{j,j+1} - J_{j-1,j}), \quad \frac{d\xi_j}{dt} = -C_{\xi,j}(J_{j,j+1} + J_{j-1,j}), \; j = 0, ...N. \quad (2.66)$$

For each $j = 0, ..., N$ coefficients $C_{P,j}$ and $C_{\xi,j}$ in (2.66) are given by (2.27), (2.28) in the cases corresponding to general mobility model (1.9) and by (2.27), (2.42) in the strong-slip case. For each $j = 1, ..., N-1$ fluxes $J_{j-1,j}$ and $J_{j,j+1}$ are given by formula (2.56) in the case of the general mobility model and by (2.65) in the case of the strong slip model. In numerical simulations of lubrication equations in the next section due to the boundary conditions (1.10) or (1.12)–(1.13) positions of the first and the last droplet in the array are fixed at the points $x = -L$ and $x = L$, respectively. Therefore, in order to complete the corresponding reduced models (2.66) we assume in them

$$J_{-1,0} := -J_{0,1}, \quad J_{N,N+1} := -J_{N-1,N}.$$

2.7 Numerical Solutions and Comparison

2.7.1 Numerical Methods

The numerical methods used in the dissertation are summarized here. For the numerical treatment of the lubrication models we used the scheme, developed in Münch et al. [1], Münch [40] and Peschka [28]. It solves the lubrication models (1.6), (1.8) and (1.2) with the boundary conditions (1.10), and the strong-slip model (1.3a)–(1.3b) with the boundary conditions (1.12)–(1.13). It is a fully implicit finite difference scheme on a general nonuniform staggered grid in space with adaptive time step. At each time step the corresponding nonlinear systems of algebraic equations is solved using Newton-Raphson method. At each Newton iteration the resulting linear system of algebraic equations is solved using effective solvers for sparse systems implemented in LAPACK library.

The numerical solutions for the reduced ODEs models (2.66) corresponding to the general mobility and strong-slip models were obtained using a fourth-order adaptive time step Runge-Kutta method in Matlab. The main difficulty was to integrate numerically coefficients (2.27), (2.28) and (2.42). The algorithm of their integration is explained already in section 2.4.

2.7.2 Numerical Solutions: Comparison and General Observations

The numerical simulations of the lubrications equations all confirmed the formation and the existence of a coarsening process for the array of quasiequilibrium droplets and supports its treatment via reduced ODE models. In this section we explain how one can iteratively solve reduced ODE models describing coarsening dynamics in large arrays of droplets and compare these results with numerical solutions of corresponding lubrication equations. We also give general for all slip cases qualitative analysis of the observed numerical results. In the next chapter we will demonstrate several important differences for the coarsening process in the strong-slip case in comparison to the slip-cases described be a type of general mobility model.

Chapter 2 Asymptotical Derivation of Reduced ODE Models

Summary of General Observations

(i) We start with explaining a strategy for simulation of coarsening dynamics in large arrays of droplets using reduced ODE models. Similarly to an algorithm introduced first in Glasner and Witelski [2], Glasner and Witelski [29] starting with an array of droplets after each subsequent coarsening event (i.e a collapse of one droplet or its collision with another) one can model the coarsening process further by reducing the dimension of the model by two. Practically we say that a collapse event occurs at a moment when pressure of one droplet increases the value $0.8 P_{max}(\varepsilon)$, where $P_{max}(\varepsilon)$ is defined in (2.1). Then we take the final pressures and positions for remaining droplets from the previous run as initial conditions for the next one. In the case of collision in Glasner and Witelski [29] was suggested to say that coarsening event occurs when the distance between two neighboring droplets becomes smaller then some $\delta = O(\varepsilon)$, i.e. when

$$(\xi_2 - A/P_2) - (\xi_1 - A/P_1) \leq \delta, \tag{2.67}$$

where $\xi_1\,\xi_2$ and $P_1\,P_2$ are positions and pressures of two colliding droplets, respectively, and A is the droplet contact angle defined in (2.44). After that we calculate the position and the pressure for the new formed droplet by formulae

$$\xi_{new} = 1/2(\xi_2 - A/P_2 + \xi_1 - A/P_1),$$
$$P_{new} = \left(\frac{1}{P_1^2} + \frac{1}{P_2^2}\right)^{-1/2}. \tag{2.68}$$

The last formula for P_{new} is based on the observation that mass of the new droplet is approximately sum of masses of colliding droplets (see Glasner and Witelski [29]). After reducing its dimension by two the ODE model still gives a good approximation for a next collapse or collision event. Example for the strong-slip case is presented in Figure 2.5 (the subsequent collapse of two droplets). We look there on the evolution of droplet pressure and compare results from lubrication and reduced ODE models. In the case of lubrication model the corresponding pressure $\Pi_\varepsilon(h) - \partial_{xx} h$ was calculated using finite-difference discretization. In the situation as on Figure 2.5 the migration of droplets is negligible, they almost do not move and collapse component dominates coarsening dynamics. Examples of subsequent modeling of collision events can be founded in the next chapter.

(ii) Next, (see Figure 2.6) we show a comparison of both pressure and position evolutions for the intermediate (1.8) and strong-slip models (1.3a)–(1.3b) for an initial profile consisting of four droplets (as in Figure 2.10). For such parameter combinations as in Figure 2.6 collapse gives again a dominant contribution to the coarsening process and collision is negligible for both models.

(iii) On Figure 2.7 we check numerically the fact that the intermediate-slip model is a limiting case of the strong-slip model as $\beta \to 0$. There we plot pressure curves for intermediate-slip model with $\beta = 1$ and show ones corresponding to the strong-slip model with a very small $\beta = 0.0001$. The initial profile is again as in Figure 2.10. One can see that if one scales the time axis in Figure 2.7 (right) by β one will have a very similar curve to those in Figure 2.7 (left).

(iv) In Figure 2.8 the validity of ODE model for the strong-slip case with flux approximation given by (2.65) is checked for different slip lengths β with fixed ϵ, Re, L. There are two general facts here that can be observed from numerical simulations. Firstly, indeed (as discussed in previous section) flux approximation (2.65) is valid only for $\beta = O(1)$. Already

2.7 Numerical Solutions and Comparison

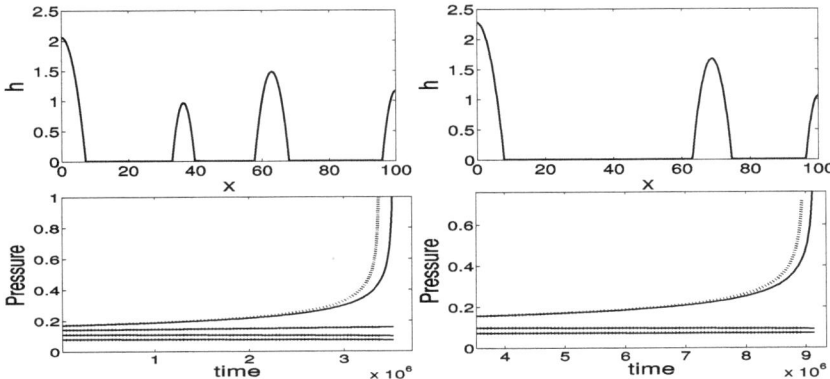

Figure 2.5: Comparison of the results of the ODE model (dotted line) and the lubrication model (solid line) for the strong-slip case with $\varepsilon = 0.01$, $L = 100$, $Re = 0$. Lower row: evolution of droplet position and pressure until collapse of the second droplet (left), and until the collapse of the fourth droplet (right). Upper row: Corresponding initial profiles for two subsequent iterations.

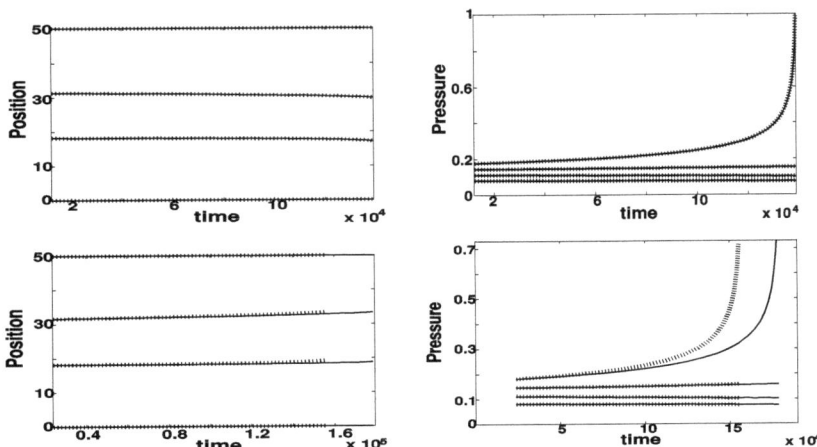

Figure 2.6: Comparisons of the ODE model and lubrication models with $\varepsilon = 0.025$, $L = 50$ in the intermediate-slip and strong-slip cases with $Re = 0$, $\beta = 1$ for the initial profile given at Fig 2.10. Collapse of the smallest (second) droplet. Lubrication model (solid line), ODE model (dotted line). Upper row: Pressure and position evolution in the intermediate-slip case. Low row: Pressure and position evolution in the strong-slip case.

Chapter 2 Asymptotical Derivation of Reduced ODE Models

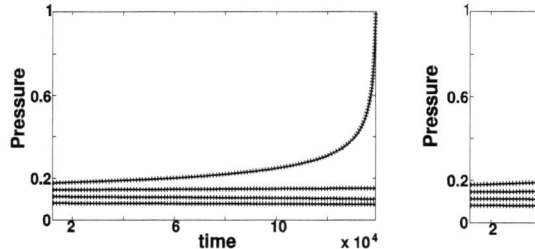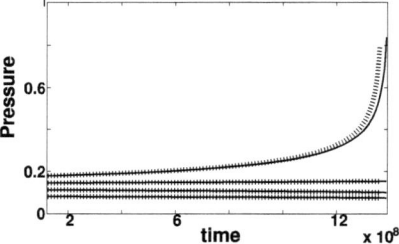

Figure 2.7: Intermediate-slip, strong-slip comparison for $\epsilon = 0.025$, $L = 50$. Intermediate-slip(left) and strong-slip with $Re = 0$, $\beta = 0.0001$ (right). Lubrication model (solid line), ODE model (dotted line).

when $\beta = 5$ this approximation is bad. Secondly, increasing β (starting from value 1 in this case) causes change in coarsening event from collapse to collision. The more precise analytical and numerical investigation for influence of slippage on coarsening mechanism can be found in the next chapter.

(v) Understanding that flux approximation (2.65) works well only for relative small β (see paragraph 2.5.2), to check independently on a kind of such approximations the validity of reduced ODE models for the whole range $0 < \beta < \infty$, we made the following simulations. Figure 2.9 illustrates the numerical simulation of the coarsening process for an array of four droplets with the initial profile as in Figure 2.10 In cases $\beta = 10$ and $\beta = 100$ we solved first lubrication model (1.3a)–(1.3b). Then for every time step $t = t^n$ of the numerical solver the constant fluxes $J_{k,k\pm1}(t^n)$, $k = 1, 2, 3$ were calculated and interpolated for time in between, resulting in the interpolated fluxes $J_{k,k\pm1}(t)$. Finally, we solved the ODE system (2.66) corresponding to the strong-slip case for the same droplet array and slippage parameter β, evaluating the flux values $J_{k,k\pm1}$ at every ODE time step using known interpolated curves for them. As a result we obtained an excellent agreement between (1.3a)–(1.3b) and its reduced model see Figure 2.9 (the cases $\beta = 10$ and $\beta = 100$). Comparison of these cases with case $\beta = 1$ (which was calculated using (2.65)) clearly indicates that a large numerical error in the case of reduced ODEs sits in the quasistationary approximation of fluxes (2.65).

(vi) Additional numerical errors for reduced ODEs come in approximation of pressure coefficient (2.27) by (2.43) and due to approximation of maximum $\hat{h}_\varepsilon^+$ and minimum $\hat{h}_\varepsilon^-$ of the droplet height profile by the leading order terms in their expansions in powers of ϵ (2.3b) and (2.3a) (during the integration of (2.28) and (2.42)). These observations suggest that numerical approximation of the lubrication models by corresponding ODEs increases when ϵ decreases. This influence of ε is justified by numerical results presented at the Figure 2.10. The solution of strong-slip ODE model is given for three different values of ε for the same initial configuration of four droplets as before.

(vii) Another interesting fact demonstrated by Figure 2.10 is as follows. When ε decreases the dominant coarsening mechanism changes from collapse to collision. For $\varepsilon = 0.1$ we have pure collapse, for $\varepsilon = 0.025$ collapse is dominant, but the motion is considerable also, for $\varepsilon = 0.01$ collision becomes dominant. This behavior is characteristic not only for the strong-slip case but also for all cases of general mobility model. For the no-slip case it was observed already before (see Bertozzi et al. [15], Glasner and Witelski [29]).

2.7 Numerical Solutions and Comparison

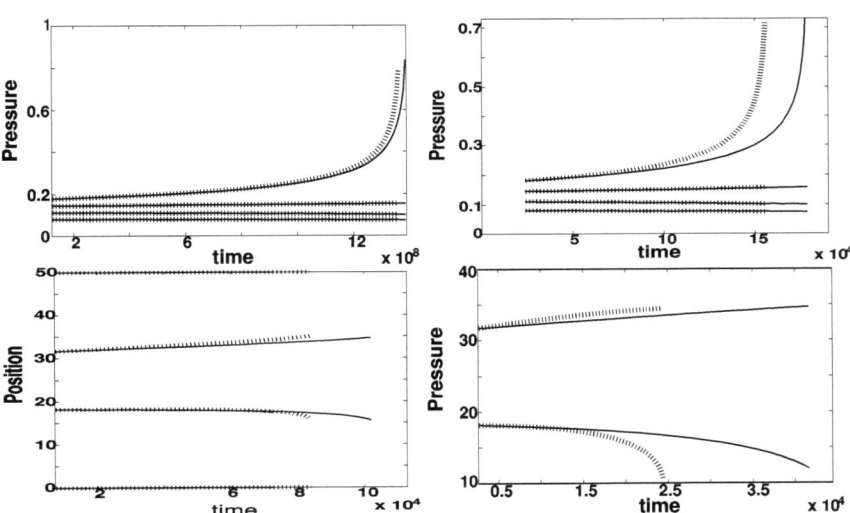

Figure 2.8: Comparison for the strong-slip case with different β. $L = 50$, $\epsilon = 0.025$ $Re = 0$. Upper row–collapse of the second droplet: $\beta = 0.0001$ (left) and $\beta = 1$ (right). Lower row–collision of the second droplet: $\beta = 2$ (left) and $\beta = 5$ (right). Lubrication model (solid line), ODE model (dotted line).

Chapter 2 Asymptotical Derivation of Reduced ODE Models

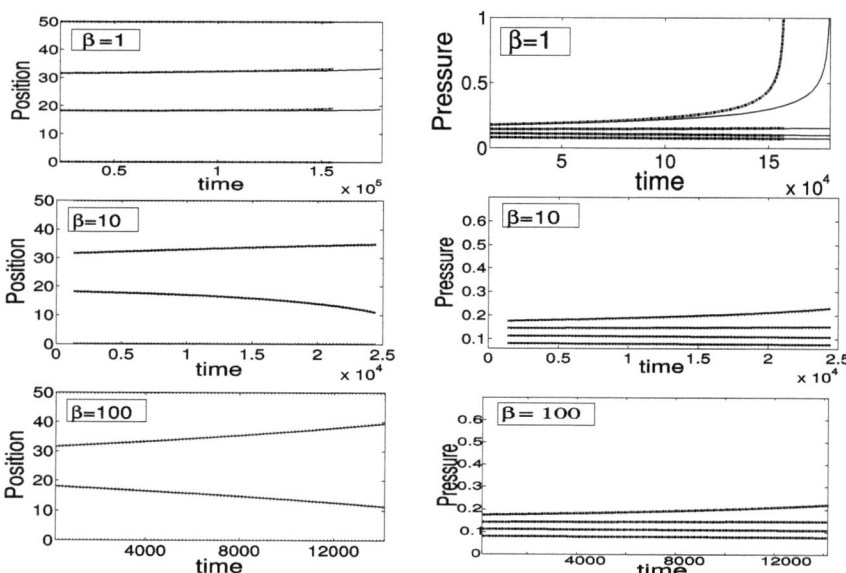

Figure 2.9: Comparison of droplet position and pressure evolution for the strong-slip case with different β. $L = 50$, $\epsilon = 0.025$ $Re = 0$. a) $\beta = 1$, b) $\beta = 10$, c) $\beta = 100$. PDE result (solid line), ODE model (dotted line).

2.7 Numerical Solutions and Comparison

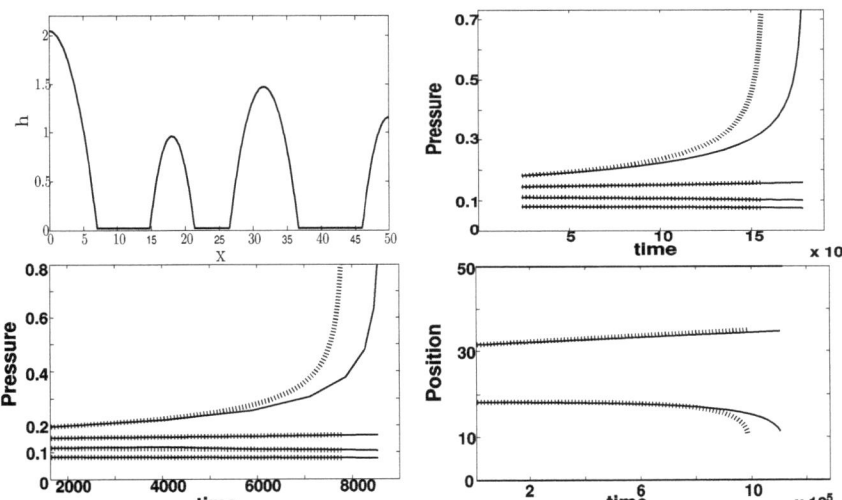

Figure 2.10: Strong-slip comparisons for different ϵ. $L = 50$, $\beta = 1$ $Re = 0$. Evolution of the smallest second droplet. Upper left: initial profile of four droplets. Pressure evolution plots (collapse is dominant) for $\epsilon = 0.1$ (upper right) and $\epsilon = 0.025$ (bottom left). Evolution of positions for $\epsilon = 0.01$ (collision is dominant, bottom right). Lubrication model (solid line), ODE model (dotted line).

Chapter 2 Asymptotical Derivation of Reduced ODE Models

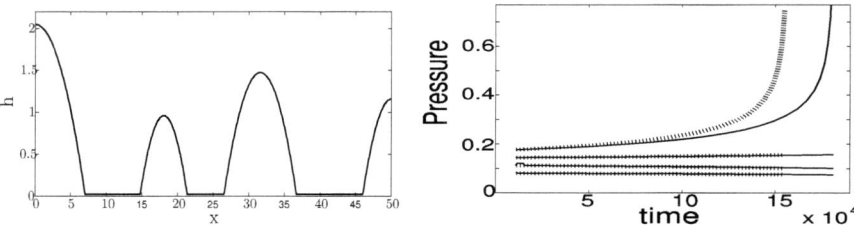

Figure 2.11: Strong-slip case for moderate $Re = 10000$, $\epsilon = 0.025$, $L = 50$ and $\beta = 1$. Initial profile of four droplets (left) and comparison of pressure evolution given by (1.3a)–(1.3b) (solid line) and the corresponding reduced model (dotted line) for the collapse of the second droplet(right). From the numerical observation the characteristic time scale of the coarsening process $\sigma \sim 10^{-5}$.

2.8 Numerical Investigation of Inertia Influence

In this dissertation we made preliminary numerical investigations of the influence of the inertia term in (1.3a)–(1.3b) and order of Re number on the coarsening dynamics in the strong-slip case. Figure 2.11 shows validity of (2.66) with (2.27), (2.42) for rather moderate Reynolds numbers. In general, determining numerically the characteristic time scale of the coarsening dynamics σ we observe, that if condition (2.29) holds then the ODE model gives a good approximation. Our numerical simulations show that for very high Re, which do not satisfy (2.29) the reduced ODE (2.40a)–(2.40b) does not approximate (1.3a)–(1.3b) anymore. When $\sigma^2 Re = O(1)$ we observe some new qualitative dynamics. The pressure inside droplets is not constant anymore and has a complicated nonlinear structure. The two components of coarsening process (collision and coarsening) still exists, but in numerical simulations one can observe oscillation behavior in droplet height profile (in particularly in ultra-thin layer) during the coarsening process. These oscillations increase, when ϵ increases (compare Figure 2.12 and Figure 2.13). Moreover, we observe a strange new effect induced by these oscillations: coming closer to collapse moment a droplet can several time disappear in the unstable UTF and then recover again to a considerable size (see example in Figure 2.13). This indicates that oscillations influence also the character of coarsening events (i.e. collapse and collision) in a new way, which is not possible for general mobility type models and the strong-slip one with $\sigma^2 Re \ll O(1)$. Although a preliminary consideration on a physical adequacy of the lubrication model (1.3a)–(1.3b) with very high Re numbers such that $\sigma^2 Re = O(1)$ is needed, nevertheless a possibility of ODE reduced model describing such a complicated coarsening dynamics could be an interesting problem for future investigations.

2.8 Numerical Investigation of Inertia Influence

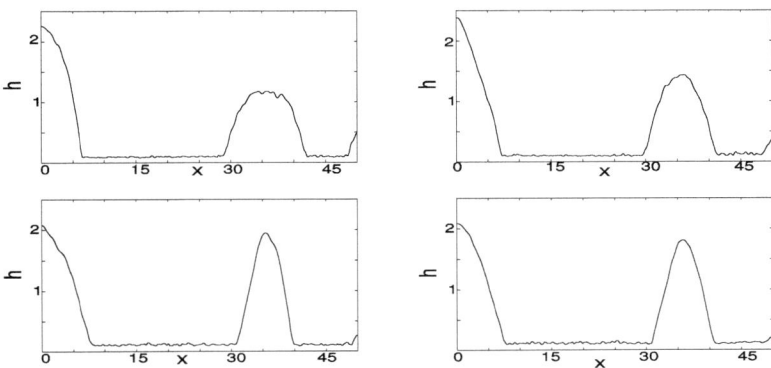

Figure 2.12: Screen-shots of the coarsening dynamics governed by the strong-slip equation with $\epsilon = 0.1$, $L = 50$, $\beta = 1$ $Re = 10^{10}$. The characteristic time scale is $\sigma = 10^{-5}$.

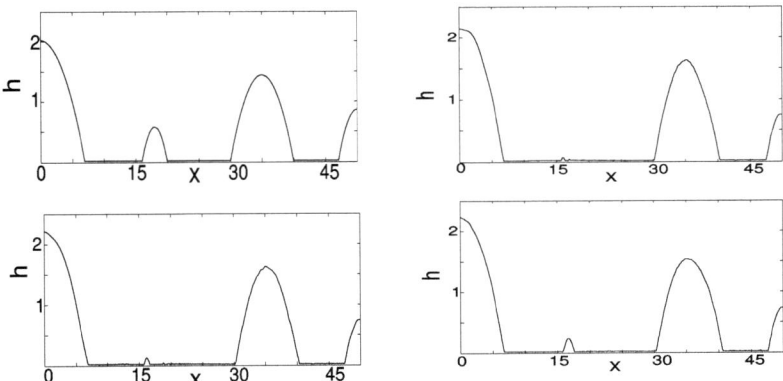

Figure 2.13: Screen-shots of the coarsening dynamics governed by the strong-slip equation with $\epsilon = 0.025$, $L = 50$, $\beta = 1$ $Re = 10^{10}$. The characteristic time scale is again $\sigma = 10^{-5}$.

Chapter 3
Slippage as a Control Parameter for Migration

3.1 Critical Value of Slippage

One of the characteristic properties of the weak-slip model (1.2) along with (1.6) and (1.8) is the following fact: the droplet migrates always opposite to the effective applied flux, see in particular the detailed discussion in Glasner et al. [3]. In the case of the j-th droplet in the array of $N+1$ droplets the effective flux applied on it is

$$J_{eff} = J_{j,j+1} + J_{j-1,j}. \tag{3.1}$$

The explanation of the property is straightforward from the migration equation (2.26) and the expression for the motion coefficient (2.28). One observes that the integrands of numerator and denominator of (2.28) are always positive, because $\hat{h}_\varepsilon(x,P) > \hat{h}_\varepsilon^-$ for all $x \in \mathbb{R}$ and hence $C_\xi > 0$. Then from (2.26) it follows that the sign of $d\xi/dt$ is always opposite to the sign of J_{eff}.

In contrast, in the case of the strong-slip model the analysis of (2.40b) and (2.42) shows, that a droplet can migrate opposite or in the same direction of the applied effective flux (3.1) depending on the value of the slip-length β. This fact is explained by the influence of the new term (2.37) in the expression for the mobility coefficient (2.42), which is connected with the presence of the Trouton viscosity term in the system (1.3a)–(1.3b).

Proposition 3.1. *There exist positive numbers $P^* > P_*$ and ε_1, K such that for any $P \in (P_*, P^*)$ and $\varepsilon \in (0, \varepsilon_1)$ there exist a unique zero $\beta = \beta_{crit}(P, \varepsilon)$ of (2.42), considered as a function of β. Moreover,*

$$\beta_{crit}(P, \varepsilon) = K\varepsilon \ln\left(\frac{2}{3\varepsilon P}\right) + o(\varepsilon) \text{ for all } P \in (P_*, P^*) \tag{3.2}$$

and $C_\xi > 0$ ($C_\xi < 0$), i.e direction of droplet migration is opposite to (in the direction of) the flux when $\beta < \beta_{crit}(P, \varepsilon)$ ($\beta > \beta_{crit}(P, \varepsilon)$).

Proof: We use here again the notation (2.23). In section 2.4 we derived asymptotics (2.54) and (2.53) for integrals

$$I_{2,m} = \int_{-\tilde{L}}^{\tilde{L}} T(\hat{h}_\varepsilon, \partial_x \hat{h}_\varepsilon, \partial_{xx} \hat{h}_\varepsilon)\left(\hat{h}_\varepsilon - \hat{h}_\varepsilon^-\right)^m dx, \ m = 0, 1.$$

and (2.52) for the integral

$$I_{1,1} = \int_{-\tilde{L}}^{\tilde{L}} \frac{\left(\hat{h}_\varepsilon - \hat{h}_\varepsilon^-\right)}{\hat{h}_\varepsilon^2} dx$$

as $\varepsilon \to 0$ holding uniformly for $P \in (P_*, P^*)$. The first asymptotics (2.54) shows that integral $I_{2,1}$ is negative for sufficiently small ε. This in turn implies that the denominator of the mobility

Chapter 3 Slippage as a Control Parameter for Migration

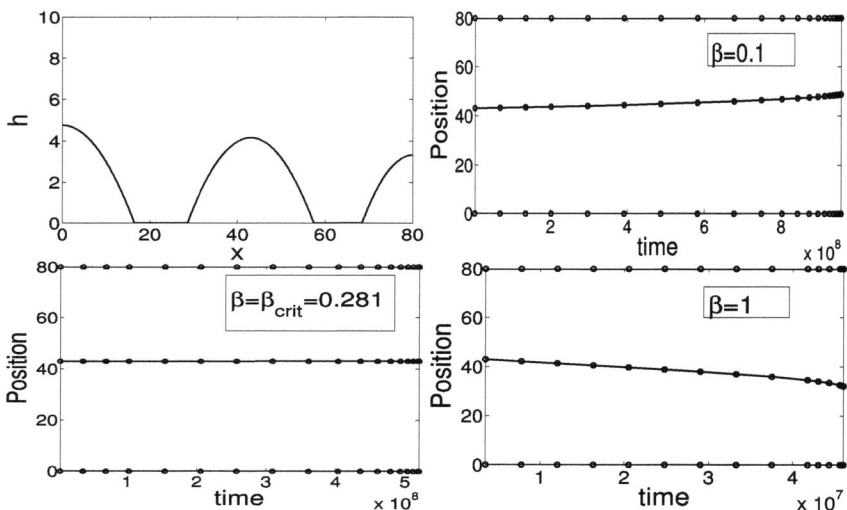

Figure 3.1: Migration of the middle droplet in the array of three droplets (upper-left) for different β. $L = 80$, $\epsilon = 0.01$, $Re = 0$, $P_2 = 0.4$.

coefficient (2.42) is positive. The second asymptotics shows that integral $I_{2,0}$ is positive for sufficiently small ε. From this and the fact that integral $I_{1,1}$ is positive, because $\hat{h}_\varepsilon > \hat{h}_\varepsilon^- > 0$, follows the existence and uniqueness of β_{crit} given as

$$\beta_{crit} := \frac{\int_{-\tilde{L}}^{\tilde{L}} \frac{\hat{h}_\varepsilon - \hat{h}_\varepsilon^-}{\hat{h}_\varepsilon^2} \, dx}{\int_{-\tilde{L}}^{\tilde{L}} T(\hat{h}_\varepsilon, \partial_x \hat{h}_\varepsilon, \partial_{xx} \hat{h}_\varepsilon) \, dx}. \tag{3.3}$$

This definition together with asymptotics (2.52) and (2.53) for integrals $I_{1,1}$ and $I_{2,m}$ imply (3.2). Moreover, from (2.42) and (3.3) it follows that $C_\xi > 0$ for $\beta < \beta_{crit}(P,\varepsilon)$ and $C_\xi < 0$ for $\beta > \beta_{crit}(P,\varepsilon)$. ∎

This migration effect is illustrated numerically in Figure 3.1. For a given initial array of three droplets β_{crit} for the middle one was calculated and then solutions of the reduced ODEs (2.66) with coefficients (2.27) and (2.42) for three different values for the slip parameter $\beta_1 < \beta_2 = \beta_{crit} < \beta_3$ were obtained. Note that in the case $\beta = \beta_{crit}$ the middle droplet almost does not move. Using the formula (2.56) and given parameters of the droplet array one can calculate that the effective flux applied on the middle droplet in all three cases is negative. So in the case $\beta < \beta_{crit}$ it moves opposite to the flux and in the case $\beta > \beta_{crit}$ in the direction of the flux.

Additionally, we recall that for the no-slip model it was shown in Glasner and Witelski [29] that collisions are typically observed for a system of at least three droplets, when two bigger droplets are attracted by a smaller droplet in between. For the strong-slip model (1.3a)–(1.3b) two-droplet collisions are typically admitted when $\beta > \beta_{crit}$. An example of a two-droplet collision is shown in Figure 3.2 with $\beta = 1$. It is a consequence of the fact that a droplet can migrate

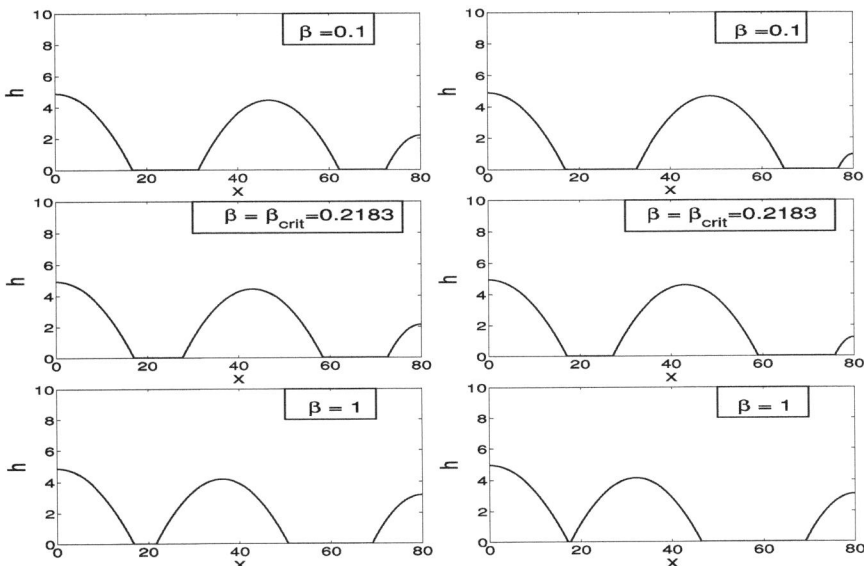

Figure 3.2: The evolution of three droplets for three different β with initial profile and parameters as in Figure. 3.1. The first and second column correspond to middle and the end of the evolution of the droplets, respectively.

in the direction of the applied flux. We remark that in the case $\beta < \beta_{crit}$ (see Figure 3.2 with $\beta = 0.1$) a two-droplet collision is not possible and the strong-slip model behaves just as the intermediate-slip one (1.8).

3.2 Coarsening Patterns for Increasing Slippage

In this section we try to analyze how the presence of critical slip-length β_{crit} given by (3.2), (3.3) influences on the coarsening dynamics on a macro-level, that is on the coarsening of large arrays of droplets. We consider a modeling example from Glasner and Witelski [29] of coarsening process of initially ten droplets (see Figures 1 and 10 in Glasner and Witelski [29]) and look how the coarsening scenario changes with slippage in comparison to the no-slip case analyzed in the latter article. Starting from an initial array of ten droplets as shown in Figure 3.3 we follow the paths of the first eight droplets in time in Figure 3.4, where we vary slippage while keeping $\varepsilon = 0.1$ fixed. First simulating intermediate-slip regime and then increasing slippage in the strong-slip regime we observe several changes in coarsening behavior, first due to the existence of different β_{crit} for each droplet in the array, and second (also as a consequence of the first fact) because migration and hence collision rates change with slippage in comparison with collapse ones.

For every droplet in the array with initial pressure P_j one can calculate $\beta_{crit,j} = \beta_{crit}(P_j)$. In the example we chose, the values for $\beta_{crit,j}$ do not differ much from each other and are approximately contained in the interval $I = [1.18, 1.3])$. When slippage is below 1.18 all droplets move opposite to the flux, but the value of their mobility coefficient in the reduced ODE (2.66)

Chapter 3 Slippage as a Control Parameter for Migration

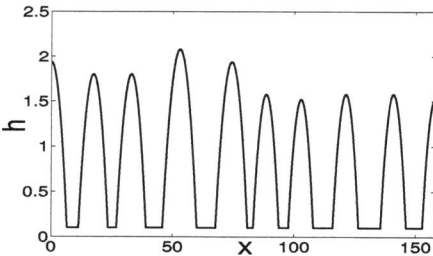

Figure 3.3: Initial profile used for all the coarsening simulations in figure 3.4

with (2.27), (2.42) tends to zero as $\beta \to 1.18$, and therefore migration rates also approach zero. This effect can be seen in Figure 3.4 for $\beta = 0.3$ and $\beta = 0.5$. In the former case the first two coarsening effects are the same as for the intermediate-slip case, namely first collapse of the 6th droplet and then collision of the 2nd and 3rd ones. But in the case $\beta = 0.5$ migration becomes so slow that the 2nd and the 3rd drops can not collide, instead the former one collapses first.

The next qualitative change in the coarsening behavior occurs when passing the critical interval I. We observe that one droplet after another change their migration direction from opposite to the flux to the same direction of the flux. For example in the case $\beta = 1.25$ smaller droplets (like the 2nd, 3rd, 6th and 7th) have already changed their migration direction, but the bigger ones not. Nevertheless, we do not observe any change in coarsening events in comparison with the case $\beta = 0.5$.

To see new events we need to increase slippage further. As a consequence the migration coefficients in the reduced model increase and hence the migration rates as well. This, together with the fact that now all droplets migrate in the same direction as the flux, considerably changes the coarsening events in our example. For example, when $\beta = 2$ we see that both the 5th and 6th droplets migrate to the left. In contrast, in the intermediate-slip case they moved to each other together with the 4th droplet, which now migrates to the right. Moreover, the 2nd and 3rd droplets do not attract each other anymore, rather, the former one collides with the first droplet and this becomes the second coarsening event in our system. The first one is collapse of the 6th droplet.

Increasing slippage even higher, here up to $\beta = 3.5$, we see that migration rates increase further, so that now the 4th and 5th droplet collide (first coarsening event) before the 1st and 2nd ones, and the 6th collapses (second coarsening event). In principle, one could increase slippage further, so that one after the other coarsening event should change from collapse to collision.

In summary we demonstrated that increasing slippage, here in the interval [0, 3.5], several different coarsening behaviors for an initial droplet array can be distinguished. They illustrate that the existence of β_{crit} influences the coarsening dynamics by changing the direction of migration and as well as migration rates. They decrease to zero, when β approaches the critical slippage interval I, and collisions become dominant after further increase of the slip-length. The latter effect is new and characteristic only for the strong-slip and free-suspended models, because as it was shown in Glasner et al. [3] in all other slip cases the migration of droplets either negligible or at most comparative with collapse component of the coarsening process. In contrast to that, in strong-slip case with sufficiently large slip-lengths migration strongly dominates the process.

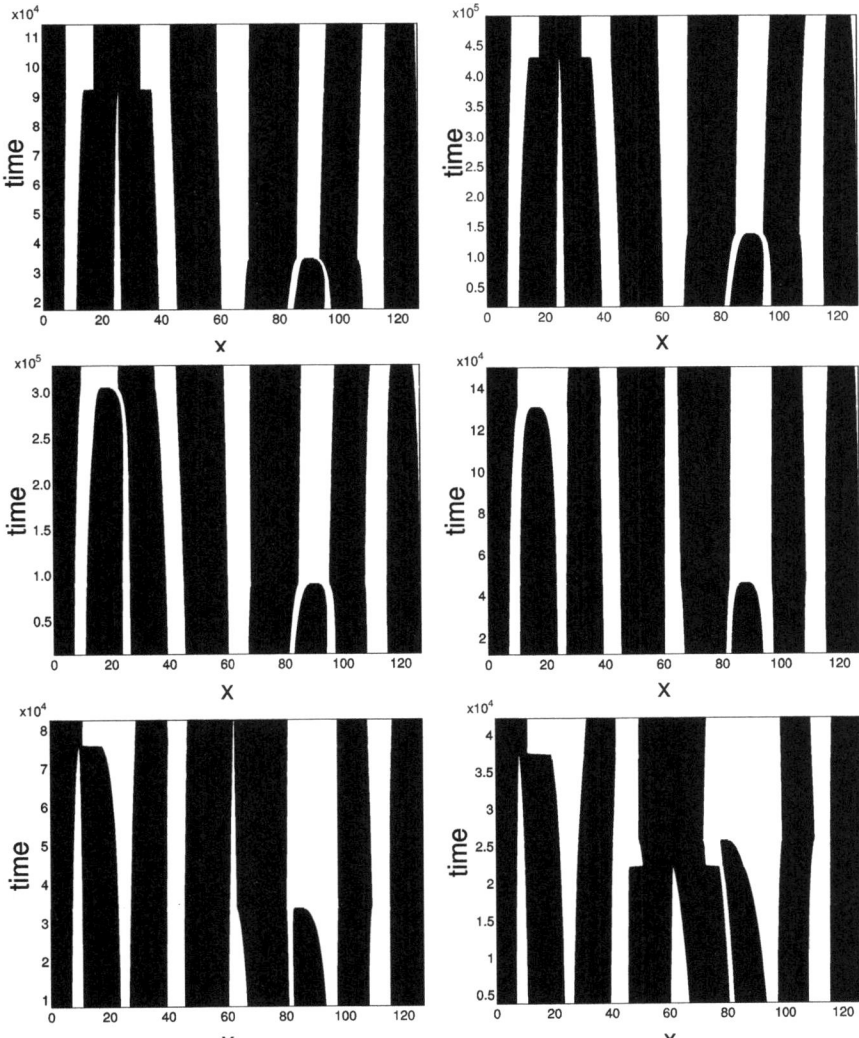

Figure 3.4: First two coarsening events in the initial array of eight droplets for six different slip values: Intermediate-slip model, strong-slip model with $\beta = 0.3$, $\beta = 0.5$, $\beta = 1.25$, $\beta = 2$, $\beta = 3.5$ (arranged from left to right and from top to bottom). In black evolution of droplet cores is plotted.

Chapter 3 Slippage as a Control Parameter for Migration

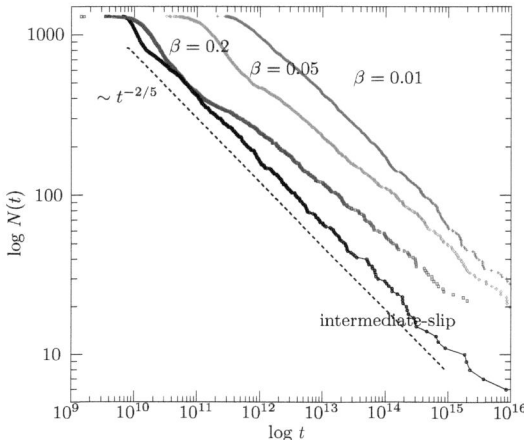

Figure 3.5: Coarsening rates for the intermediate-slip regime, strong-slip regime with $\beta = 0.2$, $\beta = 0.05$ and $\beta = 0.01$. Log-log plot for the dependence of number of droplets $N(t)$ on time.

3.3 Coarsening Rates

As was shown in previous paragraphs the value of the slip-length parameter β in the strong slip regime considerably influences the coarsening behavior of an initial system of droplets and the contribution of collision component, depending on how close this value is to the mean critical slip-length of the system (at which all droplets almost do not move and just collapse). Naturally, this fact should imply some dependence of the *collision dominated coarsening rates* on the value of the slip-length, i.e. of the coarsening rates in the systems where collisions are dominant, and collapses constitute fewer then 10% of the total amount of coarsening events. In Glasner and Witelski [2] the coarsening dynamics of initially well-separated systems of droplets which experience in general only collapse coarsening effects (so called *collapse dominated coarsening rates*), was investigated, and it was shown that the statistical number of droplets $N(t)$ in such a system changes in time according to

$$N(t) \propto t^{-2/5}. \tag{3.4}$$

Furthermore, Otto et al. [37] considered the lubrication model (1.9) with mobility term $M(h) = h$ and derived the law (3.4) on the basis of the gradient flow structure of the equation. As the equation for the evolution of droplet pressures is the same in the reduced models for all slip-regimes, it is natural to expect that the coarsening rate law for collapse dominated systems does not depend on the chosen slip regime and is always given by (3.4). Moreover, in Glasner and Witelski [29] it was shown, that for the 1D no-slip regime any initial system of droplets will coarsen according to (3.4) independent of the proportional number of collision and collapse events during the coarsening process.

In this paragraph we try to investigate collision dominated coarsening rates numerically in the intermediate and strong-slip regimes. We made simulations of coarsening rates in these cases for various initial configurations and initial numbers of droplets N_0. For this purpose we developed a Matlab program that solves effectively corresponding to above cases reduced ODE models (2.66) starting from large number of droplets around $N_0 \approx 10^3$ and modeling the coarsening process up to only several droplets remain (see Figure 3.5). As a typical initial profile we took arrays of droplets with equal small distance between them (to enforce collision dominated coarsening) and a normally distributed pressures around some $P_{mean} = 0.07$. At every iteration, in order to determine numerically a moment when coarsening event occur and the initial configuration for the next iteration after subsequent reduction of the dimension of the ODE model by two, we used stop criterion and rules for collision and collapse introduced in section 2.6.2, observation (i). In the simulations a typical dimension of ODE system is high and calculation of the mobility coefficient $C_X(P)$ is rather expensive (see details for its integration in section 2.4). Therefore, we used a multi-step variable order Adams-Bashforth-Moulton algorithm implemented in Matlab to solve ODE model at every iteration (see algorithm in Shampine and Gordon [44]). As any multi-step method it has an advantage of minimizing number of expensive evaluation of the right-hand side of the ODE system.

During simulations we always used the approximation for fluxes (2.65). It causes a certain numerical difficulty for modeling of collision events. More precisely, when the distance between two colliding droplets becomes small, i.e. near stopping criterion (2.67), the denominator of (2.56) becomes very small. This implies that one needs to solve system (2.66) in a neighborhood of a point, at which its right-hand side is singular. In this case usual error estimates (which normally use some smoothness of ODEs right-hand side) control the error of numerical solution very badly, and we needed to introduce special adaptive time step algorithm to determine the collision moment and dynamics just before it with a sufficient accuracy.

In our simulations we started first with the coarsening rates for the intermediate-slip case and found (as was expected from the results of Glasner et al. [3]) that the percentage of collisions is very small. As it is claimed in Glasner et al. [3] the intermediate-slip regime is essentially collapse dominated and hence coarsening rates for it are given again by (3.4). We then solved the reduced model (2.66) corresponding to the strong-slip lubrication equation with similar initial distributions of droplets and slip-length parameters $\beta = 0.2$, $\beta = 0.05$ and $\beta = 0.01$, respectively. Here looking at the coarsening curves in Figure 3.5 one can observe typically three time regions. After the starting one (a so called transition period) a droplet system passes to a self-similar region two, which is characterized by a straight slope of the coarsening curve. Finally, when the number of droplets becomes order of 10 a so called big size effect comes into the play and coarsening statistics becomes chaotic and not reliable.

Next, we analyzed a particular influence of slip-length β on the coarsening slope in the second self-similar time region. Firstly, we observed that the proportion of collisions is dominant and increases with slippage. Secondly, the corresponding coarsening rates have slopes $\sim t^{-1/3}$ for $\beta = 0.2$ and decrease to $\sim t^{-2/5}$ as $\beta \to 0$ as shown in Figure 3.5. This fact stays in agreement that the intermediate-slip regime is a limiting case for the strong-slip one as $\beta \to 0$. Note that for the chosen parameters ε and P_{mean} the formula (3.2) entails that the mean critical slip-length parameter $\beta_{crit} \approx 0.01$. Our chosen slip-lengths $\beta = 0.05$, $\beta = 0.2$ are beyond this value. Figure 3.5 shows, that close to β_{crit} the coarsening dynamics is as in the intermediate-slip case, but increasing the slip-length further changes continuously the coarsening slope.

Chapter 4

Formal Reduction onto an 'Approximate Invariant' Manifold

In this chapter we give a new derivation of the reduced ODE model corresponding to the no-slip lubrication equation, alternative to the asymptotical one first derived by Glasner and Witelski [2] (and then generalized for other lubrication models in Chapter I), but still formal. This derivation was inspired by a recent paper of Mielke and Zelik [4], in which authors prove rigorously a center manifold reduction theorem for a general class of semilinear parabolic equations possessing a so called *multipulse solutions*, i.e. solutions that in some sense are very close for all times to certain combinations of finite or even infinite number of stationary solutions (so called pulses) parameterized by a discrete set of parameters. For example in the case of lubrication equations one could understand stationary solution $\hat{h}_\varepsilon(x, P)$ on $\mathbb{R}$ under such a pulse. Trying to apply and modify the approach of Mielke and Zelik [4] to the no-slip lubrication equation we saw several differences and difficulties in comparison with their case (see section 4.5). Nevertheless, following main steps of the approach of Mielke and Zelik [4] and adopting several constructions to the no-slip lubrication equation, we arrived formally to a reduced ODE model, which governs the evolution of the above discrete set of parameters. Further, we compared this ODE model with the one derived by Glasner and Witelski [2].

We present this alternative derivation here for several reasons. Firstly, the invariant manifold based approach given below is quite different to that one of Glasner and Witelski [2] and gives a nice geometric interpretation for the reduced dynamics. Secondly, results of Mielke and Zelik [4] and our application of them in this chapter suggest that this derivation can be made rigorous after solving difficulties arising in the case of lubrication equation (see section 4.5).

The structure of this chapter is as follows. In section 4.1 we introduce an 'approximate invariant' manifold $\mathbb{P}_\varepsilon$, prove that every point $\mathbf{m}$ of it is 'almost stationary' with respect to the no-slip equation and define a special projection operator $P_\mathbf{m}$ on $\mathbb{P}_\varepsilon$. In section 4.2 we prove that in a sufficiently small neighborhood of the 'approximate invariant' manifold every solution $h(\cdot, t)$ of (1.6) can be decomposed into the sum of some point $\mathbf{m}(t)$ on the manifold and a reminder function $v(t)$, which is 'orthogonal' to the manifold, i.e $P_\mathbf{m} v(t) = 0$ for $t > 0$. Next, we decompose (1.6) into a system of two equations: an ODE which describes an evolution on the 'approximate invariant' manifold for $\mathbf{m}(t)$ and a quasilinear equation for the reminder $v(t)$. Up to this moment we proceed rigorously. Nevertheless, for the reasons explained in section 4.5, further we make a formal assumption on the smallness of remainder function $v(t)$ and obtain by this a formal leading order equation for $\mathbf{m}(t)$ on $\mathbb{P}_\varepsilon$ in section 4.3, which gives us a reduced ODE model. Then we compare it with reduced ODE model (2.66) corresponding to the no-slip lubrication equation in section 4.4 pointing out a good agreement between them and certain advantages of the just derived one. In section 4.5 we discuss a possible approach to make the above derivation rigorous and introduce a so called spectral problem, analytical treatment of which is the subject of the next chapter.

Chapter 4 Formal Reduction onto an 'Approximate Invariant' Manifold

4.1 'Approximate Invariant' Manifold: Definition and Properties

Let us write the no-slip lubrication model (1.6) in the form

$$\partial_t h + \mathbb{F}_\varepsilon(h) = 0 \tag{4.1}$$

and define the corresponding to it quasilinear elliptic operator as

$$\mathbb{F}_\varepsilon(h) := \partial_x\Big(h^3 \partial_x (\partial_{xx} h - \Pi_\varepsilon(h))\Big). \tag{4.2}$$

As before we put on (4.1) Neumann boundary conditions on a fixed interval $(-L, L)$:

$$\partial_{xxx} h = 0, \quad \text{and} \quad \partial_x h = 0 \quad \text{at} \quad x = \pm L. \tag{4.3}$$

Let $\hat{h}_\varepsilon(x, P)$ be the stationary solution to (4.1) on $\mathbb{R}$ defined in Theorem 2.1. Note that a shifted function $\hat{h}_\varepsilon(x - \xi, P)$ for every $\xi \in \mathbb{R}$ gives also a solution to (4.1) on $\mathbb{R}$. Recall that by Proposition 2.2 there exists positive numbers d and $P^* > P_*$ such that for $|x| > d$, $P \in (P_*, P^*)$ and sufficiently small $\varepsilon > 0$ estimates (2.8a)–(2.8c) hold

Let us next define a set $\mathbb{B}_\varepsilon \subset \mathbb{R}^{2N}$ as

$$\mathbb{B}_\varepsilon = \Big\{ \mathbf{s} = (P_0, P_1, ..., P_N, \xi_1, \xi_2 ..., \xi_{N-1}) \in \mathbb{R}^{2N} : P_j \in (P_*, P^*), j = 0, ..., N;$$
$$-L < \xi_1 < ... < \xi_{N-1} < L;\ \xi_i - \xi_{i-1} - 4d > 2\sqrt{\varepsilon},\ i = 1, ..., N \Big\}, \tag{4.4}$$

where we assumed $\xi_0 := -L$ and $\xi_N := L$. Throughout the whole chapter we fix positive numbers

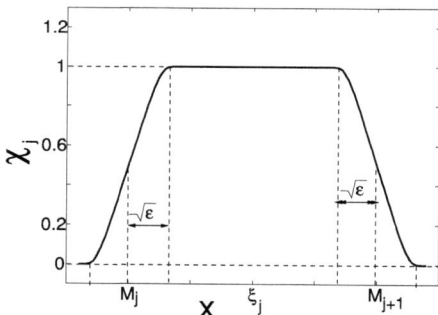

Figure 4.1: Plot of function $\chi_j(\mathbf{s})(x)$.

ε_1 and L so that set $\mathbb{B}_\varepsilon$ is not empty for all $\varepsilon \in (0, \varepsilon_1)$. The boundary of the open set $\mathbb{B}_\varepsilon$ in $\mathbb{R}^{2N}$ topology is given by

$$\partial \mathbb{B}_\varepsilon = \Big\{ \mathbf{s} \in \mathbb{R}^{2N} : \exists j \in \{0, ..., N\} : P_j = P_* \Big\}$$
$$\cup \Big\{ \mathbf{s} \in \mathbb{R}^{2N} : \exists j \in \{0, ..., N\} : P_j = P^* \Big\}$$
$$\cup \Big\{ \mathbf{s} \in \mathbb{R}^{2N} : \exists i \in \{1, ..., N\},\ \xi_i - \xi_{i-1} - 4d = 2\sqrt{\varepsilon} \Big\}.$$

4.1 'Approximate Invariant' Manifold: Definition and Properties

Let us define for each $\mathbf{s} \in \mathbb{B}_\varepsilon$ and $j \in \{1, ..., N\}$ points

$$M_j := \frac{\xi_j + \xi_{j-1}}{2} \tag{4.5}$$

and functions $\chi, \chi_j(\mathbf{s}) \in C^\infty(\mathbb{R})$ (see Figure 4.1) as

$$\chi(x) := \begin{cases} 0, & x \leq -\sqrt{\varepsilon} \\ \frac{1}{2}\left(1 + \tanh\left(\tan\left(\frac{\pi}{2\sqrt{\varepsilon}}x\right)\right)\right), & -\sqrt{\varepsilon} < x < \sqrt{\varepsilon} \\ 1, & x \geq \sqrt{\varepsilon} \end{cases}, \tag{4.6}$$

$$\chi_j(\mathbf{s})(x) = \chi_j(\xi_{j-1}, \xi_j, \xi_{j+1}, P_{j-1}, P_j, P_{j+1}, x) :=$$
$$:= \begin{cases} \chi(x - M_j), & x < M_j + \sqrt{\varepsilon} \\ 1, & M_j + \sqrt{\varepsilon} \leq x \leq M_{j+1} - \sqrt{\varepsilon} \\ 1 - \chi(x - M_{j+1}), & x > M_{j+1} - \sqrt{\varepsilon} \end{cases}, \; j = 1, ..., N-1;$$

$$\chi_0(\mathbf{s})(x) = \chi_0(\xi_1, P_0, P_1, x) :=$$
$$:= \begin{cases} 1, & 0 \leq x \leq M_1 - \sqrt{\varepsilon} \\ 1 - \chi(x - M_1), & x > M_1 - \sqrt{\varepsilon} \end{cases};$$

$$\chi_N(\mathbf{s})(x) = \chi_0(\xi_{N-1}, P_{N-1}, P_N, x) :=$$
$$:= \begin{cases} \chi(x - M_N), & x \leq M_N + \sqrt{\varepsilon} \\ 1, & x > M_N + \sqrt{\varepsilon} \end{cases}. \tag{4.7}$$

One can see that for all $x \in [0, L]$ and $\mathbf{s} \in \mathbb{B}_\varepsilon$ it holds $\sum_{j=0}^{N} \chi_j(\mathbf{s})(x) \equiv 1$. Define next a mapping $\mathbf{m}_\varepsilon : \mathbb{B}_\varepsilon \to L^\infty(-L, L)$, which maps a point $\mathbf{s} \in \mathbb{B}_\varepsilon$ to a function $\mathbf{m}_\varepsilon(\mathbf{s}) \in C^\infty(-L, L)$ satisfying boundary conditions (4.3) as follows:

$$\forall \mathbf{s} \in \mathbb{B}_\varepsilon \; \mathbf{m}_\varepsilon(\mathbf{s})(x) := \sum_{j=0}^{N} \chi_j(\mathbf{s})(x) \hat{h}_\varepsilon(x - \xi_j, P_j), \tag{4.8}$$

where again $\xi_0 = -L$, $\xi_N = L$. The image of $\mathbf{m}_\varepsilon$ defines a smooth $2N$-dimensional submanifold in L^∞, which we denote as $\mathbb{P}_\varepsilon$. Like in Mielke and Zelik [4] we define a boundary of $\mathbb{P}_\varepsilon$ as $\partial \mathbb{P}_\varepsilon := \mathbf{m}_\varepsilon(\partial \mathbb{B}_\varepsilon)$. From (4.8) it follows that every point $\mathbf{m}(\mathbf{s}) \in \mathbb{P}_\varepsilon$ is a composition of $N+1$ stationary solutions to the lubrication model (4.1). Following to Mielke and Zelik [4] we call such a composition as a *multi-droplet* or a *multi-pulse structure* (see example in Figure 4.2). The mapping $\mathbf{m}_\varepsilon$ is an diffeomorphism between $\mathbb{B}_\varepsilon$ and $\mathbb{P}_\varepsilon$, and therefore below in this chapter for each $\mathbf{m} \in \mathbb{P}_\varepsilon$ we associate a unique $\mathbf{s} \in \mathbb{B}_\varepsilon$ such that $\mathbf{m}_\varepsilon(\mathbf{s}) := \mathbf{m}$.

The tangent space $\mathbb{T}_\mathbf{m}\mathbb{P}_\varepsilon$ of manifold $\mathbb{P}_\varepsilon$ at a point $\mathbf{m} \in \mathbb{P}_\varepsilon$ is given by span of functions $\{\phi_0(\mathbf{s}), \phi_1(\mathbf{s}), ..., \phi_{2N-1}(\mathbf{s})\}$, where $\phi_j(\mathbf{s}) \in C_c^\infty(-L, L)$ are defined as follows:

$$\phi_j(\mathbf{s}) := \frac{\partial \mathbf{m}_\varepsilon(\mathbf{s})}{\partial P_j} \quad \text{for} \; j = 0, ..., N;$$
$$\phi_{N+j}(\mathbf{s}) := \frac{\partial \mathbf{m}_\varepsilon(\mathbf{s})}{\partial \xi_j} \quad \text{for} \; j = 1, ..., N-1. \tag{4.9}$$

Using definitions (4.5)–(4.8) one can see that for $j \in \{1, ..., N-1\}$ functions $\phi_j(\mathbf{s})(x)$ and

Chapter 4 Formal Reduction onto an 'Approximate Invariant' Manifold

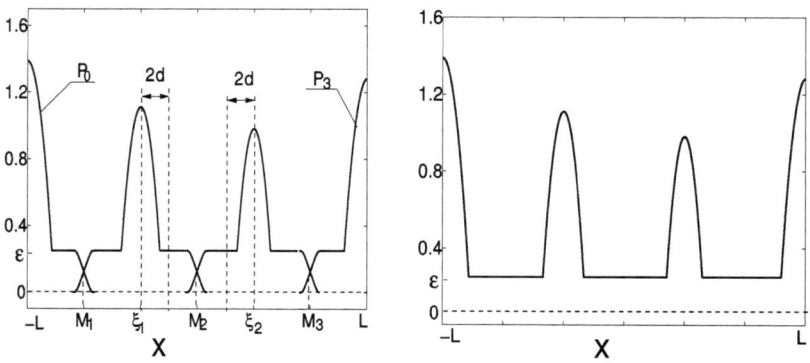

Figure 4.2: Example of a multi-droplet structure, four truncated pulses $\chi_j(\mathbf{s})(x)\hat{h}_\varepsilon(x - \xi_j, P_j)$, $j = 0, 1, 2, 3$ (left) and their sum $\mathbf{m}_\varepsilon(\mathbf{s})(x)$ (right).

$\phi_{N+j}(\mathbf{s})(x)$ have a compact support on an interval

$$I_j := (M_j - \sqrt{\varepsilon}, \ M_{j+1} + \sqrt{\varepsilon}) \qquad (4.10)$$

and can be represented as:

$$\phi_j(\mathbf{s})(x) = \begin{cases} \dfrac{\partial \hat{h}_\varepsilon(x - \xi_j, P_j)}{\partial P}\chi(x - M_j) & , \ x < M_j + \sqrt{\varepsilon} \\[1em] \dfrac{\partial \hat{h}_\varepsilon(x - \xi_j, P_j)}{\partial P} & , \ x \in [M_j + \sqrt{\varepsilon}, \ M_{j+1} - \sqrt{\varepsilon}]; \\[1em] \dfrac{\partial \hat{h}_\varepsilon(x - \xi_j, P_j)}{\partial P}(1 - \chi(x - M_{j+1})) & , \ x > M_{j+1} - \sqrt{\varepsilon} \end{cases}$$

$$\phi_{N+j}(\mathbf{s})(x) = \begin{cases} \dfrac{1}{2}\chi'(x - M_j)\left(\hat{h}_\varepsilon(x - \xi_{j-1}, P_{j-1}) - \hat{h}_\varepsilon(x - \xi_j, P_j)\right) - \\[0.5em] \quad - \dfrac{\partial \hat{h}_\varepsilon(x - \xi_j, P_j)}{\partial x}\chi(x - M_j), & x < M_j + \sqrt{\varepsilon} \\[1em] -\dfrac{\partial \hat{h}_\varepsilon(x - \xi_j, P_j)}{\partial x}, & x \in [M_j + \sqrt{\varepsilon}, \ M_{j+1} - \sqrt{\varepsilon}] \\[1em] \dfrac{1}{2}\chi'(x - M_{j+1})\left(\hat{h}_\varepsilon(x - \xi_j, P_j) - \hat{h}_\varepsilon(x - \xi_{j+1}, P_{j+1})\right) - \\[0.5em] \quad - \dfrac{\partial \hat{h}_\varepsilon(x - \xi_j, P_j)}{\partial x}(1 - \chi(x - M_{j+1})), & x > M_{j+1} - \sqrt{\varepsilon} \end{cases}$$

4.1 'Approximate Invariant' Manifold: Definition and Properties

The remaining two functions have a representation:

$$\phi_0(\mathbf{s})(x) = \begin{cases} \dfrac{\partial \hat{h}_\varepsilon(x+L, P_0)}{\partial P} & , \ x \in [0, \ M_1 - \sqrt{\varepsilon}] \\[1em] \dfrac{\partial h_\varepsilon(x+L, P_0)}{\partial P}(1 - \chi(x - M_1)) & , \ x > M_1 - \sqrt{\varepsilon}; \end{cases}$$

$$\phi_N(\mathbf{s})(x) = \begin{cases} \dfrac{\partial \hat{h}_\varepsilon(x-L, P_N)}{\partial P}\chi(x - M_N) & , \ x < M_N + \sqrt{\varepsilon} \\[1em] \dfrac{\partial \hat{h}_\varepsilon(x-L, P_N)}{\partial P} & , \ x \in [M_N + \sqrt{\varepsilon}, \ L]. \end{cases}$$

The next proposition shows that the right-hand side (4.2) of the lubrication equation (4.1) is small on the manifold $\mathbb{P}_\varepsilon$ (due to the fact that it is formed by compositions of stationary solutions). This explains why we call $\mathbb{P}_\varepsilon$ as 'approximate stationary' or 'approximate invariant'.

Proposition 4.1. *For every* $\mathbf{m} \in \mathbb{P}_\varepsilon$ *and sufficiently small* $\varepsilon > 0$ *one has*

$$\left\|\mathbb{F}_\varepsilon(\mathbf{m})\right\|_{L^\infty(-L, L)} \leq K\varepsilon^{3/2},$$

where constant $K > 0$ *does not depend on* $\mathbf{m}$, ε.

Proof: Let us fix below any $\mathbf{m} \in \mathbb{P}_\varepsilon$. Due to definitions (4.7)–(4.8) for all $x \in [0, \ M_1 - \sqrt{\varepsilon}] \cup [M_N + \sqrt{\varepsilon}, \ L]$ one has $\mathbb{F}_\varepsilon(\mathbf{m})(x) \equiv 0$. Let us estimate $\mathbb{F}_\varepsilon(\mathbf{m})(x)$ on the interval I_j from (4.10) for every $j \in \{1, ..., N-1\}$. Due to (4.7)–(4.8) one has a representation:

$$\mathbb{F}_\varepsilon(\mathbf{m})(x) = \begin{cases} \mathbb{F}_\varepsilon\Big((1 - \chi(x - M_j))\hat{h}_\varepsilon(x - \xi_{j-1}, P_{j-1})+ \\ + \chi(x - M_j)\hat{h}_\varepsilon(x - \xi_j, P_j)\Big), & x \in [M_j - \sqrt{\varepsilon}, \ M_j + \sqrt{\varepsilon}] \\[1em] 0, & x \in [M_j + \sqrt{\varepsilon}, \ M_{j+1} - \sqrt{\varepsilon}]. \\[1em] \mathbb{F}_\varepsilon\Big((1 - \chi(x - M_{j+1}))\hat{h}_\varepsilon(x - \xi_j, P_j)+ \\ + \chi(x - M_{j+1})\hat{h}_\varepsilon(x - \xi_{j+1}, P_{j+1})\Big), & x \in [M_{j+1} - \sqrt{\varepsilon}, \ M_{j+1} + \sqrt{\varepsilon}] \end{cases}$$
(4.11)

Let us first estimate $\mathbb{F}_\varepsilon(\mathbf{m})(x)$ for $x \in [M_j - \sqrt{\varepsilon}, \ M_j + \sqrt{\varepsilon}]$. Due to asymptotics (2.3a) and definitions (4.6), (4.8) for sufficiently small $\varepsilon > 0$ it holds

$$\varepsilon \leq \min\left\{\hat{h}_\varepsilon(x - \xi_j, P_j), \ \hat{h}_\varepsilon(x - \xi_{j-1}, P_{j-1})\right\} \leq |\mathbf{m}(x)| \leq$$
$$\leq \max\left\{\hat{h}_\varepsilon(x - \xi_j, P_j), \ \hat{h}_\varepsilon(x - \xi_{j-1}, P_{j-1})\right\} \leq \varepsilon + 2P^*\varepsilon^2, \quad (4.12)$$

where min and max are taken in $x \in [M_j - \sqrt{\varepsilon}, \ M_j + \sqrt{\varepsilon}]$. Therefore, for such x and sufficiently small $\varepsilon > 0$ one obtains

$$|\Pi_\varepsilon(\mathbf{m}(x))| = \left|\varepsilon^{-1}\left(\left(\frac{\varepsilon}{\mathbf{m}}\right)^3 - \left(\frac{\varepsilon}{\mathbf{m}}\right)^4\right)\right| \leq$$
$$\leq \left|\varepsilon^{-1}\left(1 - \left(\frac{1}{1+2P^*\varepsilon}\right)^4\right)\right| \leq K_0.$$

In the same manner one obtains

$$|\Pi'_\varepsilon(\mathbf{m}(x))| \leq K_1/\varepsilon^2 \quad \text{and} \quad |\Pi''_\varepsilon(\mathbf{m}(x))| \leq K_2/\varepsilon^3,$$

where constants K_i, $i = 0, 1, 2$ do not depend on $\mathbf{m} \in \mathbb{P}_\varepsilon$, $\varepsilon > 0$ and $x \in [M_j - \sqrt{\varepsilon},\ M_j + \sqrt{\varepsilon}]$. Using definition (4.6) one obtains

$$\left|\frac{d^k\chi}{dx^k}\right| \leq \left(\frac{\pi}{2\sqrt{\varepsilon}}\right)^k, \text{ for } k \in \mathbb{N}_0 \text{ uniformly in } x \in \mathbb{R}. \tag{4.13}$$

By estimate (2.8b) and definition (4.4)

$$\left|\frac{\partial^k \hat{h}_\varepsilon(x - \xi_j, P_j)}{\partial x^k}\right| \leq \frac{C_0}{\varepsilon^k} \exp\left(-\frac{d}{\sqrt{2}\varepsilon}\right)$$

for all $\mathbf{s} \in \mathbb{B}_\varepsilon$, $j = 0, ..., N$, $x \in [M_j - \sqrt{\varepsilon}, M_j + \sqrt{\varepsilon}]$ and $k = 1, ..., 4$. Therefore, using also (4.12) one obtains

$$\left|\frac{\partial^k \mathbf{m}(x)}{\partial x^k}\right| \leq K_3\, \varepsilon^{1-k/2},\ \ k = 0,\ 1, ..., 4,$$

where constant $K_3 > 0$ does not depend on $\mathbf{m} \in \mathbb{P}_\varepsilon$, $\varepsilon > 0$ and $x \in [M_j - \sqrt{\varepsilon},\ M_j + \sqrt{\varepsilon}]$. Finally, using last five estimates one obtains for all $x \in [M_j - \sqrt{\varepsilon},\ M_j + \sqrt{\varepsilon}]$

$$|\mathcal{F}_\varepsilon(\mathbf{m})(x)| \leq |\mathbf{m}^3 \mathbf{m}_{xxxx}| + |\mathbf{m}^3 \Pi'_\varepsilon(\mathbf{m})\mathbf{m}_{xx}| + |\mathbf{m}^3 \Pi''_\varepsilon(\mathbf{m})\mathbf{m}_x^2|$$
$$+ |3\mathbf{m}^2\mathbf{m}_x\mathbf{m}_{xxx}| + |3\mathbf{m}^2\mathbf{m}_x\Pi'_\varepsilon(\mathbf{m})\mathbf{m}_x| \leq K_5\, \varepsilon^{3/2}.$$

In the same manner analogous estimate on $|\mathcal{F}_\varepsilon(\mathbf{m})|$ can be obtained for $x \in [M_{j+1} - \sqrt{\varepsilon},\ M_{j+1} + \sqrt{\varepsilon}]$, and therefore using (4.11) one ends up with

$$\left\|\mathcal{F}_\varepsilon(\mathbf{m})\right\|_{L^\infty(I_j)} \leq K\varepsilon^{3/2} \text{ for every } j \in \{1, ..., N-1\}.$$

This finishes the proof. ∎

Next, for each $\mathbf{m} \in \mathbb{P}_\varepsilon$ we would like to define a projection on $\mathbb{T}_\mathbf{m}\mathbb{P}_\varepsilon$ in $L^\infty(-L, L)$ using a so called "adjoint functions" $\psi_j(\mathbf{s}) \in C_c^\infty(-L, L)$, $j = 0, ..., 2N - 1$. Namely, define

$$\psi_j(\mathbf{s})(x) := C_j(\mathbf{s})\chi_j(\mathbf{s})(x),\ j = 0, ..., N;$$
$$\psi_{N+j}(\mathbf{s})(x) := C_{N+j}(\mathbf{s})\chi_j(\mathbf{s})(x) \int_{\xi_j}^x \frac{\hat{h}_\varepsilon(s - \xi_j, P_j) - \hat{h}_\varepsilon^-(P_j)}{\hat{h}_\varepsilon(s - \xi_j, P_j)^3}\, ds,\ j = 1, ..., N-1,$$

$$\tag{4.14}$$

where we denoted

$$C_j(\mathbf{s}) := 1/\int_{M_j+\sqrt{\varepsilon}}^{M_{j+1}-\sqrt{\varepsilon}} \frac{\partial \hat{h}_\varepsilon(x-\xi_j, P_j)}{\partial P}\, dx,$$

$$C_{N+j}(\mathbf{s}) := 1/\int_{M_j+\sqrt{\varepsilon}}^{M_{j+1}-\sqrt{\varepsilon}} \frac{\left(\hat{h}_\varepsilon(x-\xi_j, P_j) - \hat{h}_\varepsilon^-(P_j)\right)^2}{\hat{h}_\varepsilon(x-\xi_j, P_j)^3}\, dx,\ \text{for}\ j=1,...,N-1 \qquad (4.15)$$

and

$$C_0(\mathbf{s}) := 1/\int_{-L}^{M_1-\sqrt{\varepsilon}} \frac{\partial \hat{h}_\varepsilon(x+L, P_0)}{\partial P}\, dx,$$

$$C_N(\mathbf{s}) := 1/\int_{M_N+\sqrt{\varepsilon}}^{L} \frac{\partial \hat{h}_\varepsilon(x-L, P_N)}{\partial P}\, dx.$$

Again for $j \in \{1,...,N-1\}$ functions $\psi_j(\mathbf{s})(x)$ and $\psi_{N+j}(\mathbf{s})(x)$ have compact supports on interval I_j given by (4.10).

Remark 4.2. Recall that formal adjoint $\mathbb{F}_\varepsilon{}'(\mathbf{m})^*$ to the operator obtained by differentiation of $\mathbb{F}_\varepsilon(\mathbf{m})$ at a point $\mathbf{m} \in \mathbb{P}_\varepsilon$, due to definitions (4.2), (4.8), acts as:

$$\mathbb{F}_\varepsilon{}'(\mathbf{m})^*[\psi(\mathbf{s})](x) = \left(\Pi'_\varepsilon\left(\hat{h}_\varepsilon(x-\xi_j, P_j)\right) - \partial_{xx}\right)\left[\partial_x\left(\hat{h}_\varepsilon(x-\xi_j, P_j)^3 \psi(x)\right)\right]$$

for $x \in [M_j + \sqrt{\varepsilon}, M_{j+1} - \sqrt{\varepsilon}]$. From this it follows that $\mathbb{F}_\varepsilon{}'(\mathbf{m})^*[\psi_j(\mathbf{s})](x) \equiv 0$ for $x \in [M_j + \sqrt{\varepsilon}, M_{j+1} - \sqrt{\varepsilon}]$. Therefore, we call functions $\psi_j(\mathbf{s})$ as 'adjoint'. Similar functions we have used already during the asymptotical derivation of reduced models in section 2.3. ∎

Before defining a projection on $\mathbb{T}_\mathbf{m}\mathbb{P}_\varepsilon$ we prove two helpful propositions.

Proposition 4.3. *There exists a positive number $K > 0$ such that for all $\mathbf{m} \in \mathbb{P}_\varepsilon$, sufficiently small $\varepsilon > 0$ and $j, k \in \{0,...,2N-1\}$ one has*

$$\left|(\psi_j(\mathbf{s}),\, \phi_k(\mathbf{s})) - \delta_{j,k}\right| \le K\varepsilon^{3/2}, \qquad (4.16)$$

where $(\cdot,\, \cdot)$ denotes the standard inner product in $L^2(-L, L)$.

Proof:
a) Let us first consider $(\psi_j(\mathbf{s}),\, \phi_j(\mathbf{s}))$ for $j \in \{1,...,N-1\}$. By definitions (4.9) and (4.14) one has:

$$\frac{(\psi_j,\, \phi_j)}{C_j(\mathbf{s})} = \int_{M_j+\sqrt{\varepsilon}}^{M_{j+1}-\sqrt{\varepsilon}} \frac{\partial \hat{h}_\varepsilon(x-\xi_j, P_j)}{\partial P}\, dx +$$
$$+ \int_{M_j-\sqrt{\varepsilon}}^{M_j+\sqrt{\varepsilon}} \left(\frac{\partial \hat{h}_\varepsilon(x-\xi_j, P_j)}{\partial P}\chi(x-M_j)\right)\chi(x-M_j)\, dx +$$
$$+ \int_{M_{j+1}-\sqrt{\varepsilon}}^{M_{j+1}+\sqrt{\varepsilon}} \left(\frac{\partial \hat{h}_\varepsilon(x-\xi_j, P_j)}{\partial P}\chi(x-M_{j+1})\right)(1-\chi(x-M_{j+1}))\, dx.$$

By (2.8c) and (4.4) there exists a positive number K_0 such that for all $\mathbf{s} \in \mathbb{B}_\varepsilon$, $x \in [\xi_{j-1}+2d,\, \xi_j - 2d]$ with $j=1,...,N$ and sufficiently small $\varepsilon > 0$ it holds

$$\frac{\partial \hat{h}_\varepsilon(x-\xi_j, P_j)}{\partial P} \le K_0\varepsilon. \qquad (4.17)$$

Chapter 4 Formal Reduction onto an 'Approximate Invariant' Manifold

From the last two expressions and estimate (4.13) one obtains:

$$|(\psi_j(\mathbf{s}),\ \phi_j(\mathbf{s})) - 1| \leq K_1\, \varepsilon^{3/2} C_j(\mathbf{s}).$$

By definition (4.15) $C_j(\mathbf{s})$ is bounded uniformly in $\mathbf{s} \in \mathbb{B}_\varepsilon$ and $j = 0, ..., N$. Therefore, estimate (4.16) for this case follows.

b) Let us consider $(\psi_j(\mathbf{s}),\ \phi_{N+j}(\mathbf{s}))$ for $j \in \{1, ..., N-1\}$. By definitions (4.5), (4.6) and (4.9), (4.14) one has:

$$\frac{(\psi_j,\ \phi_{N+j})}{C_j(\mathbf{s})} = -\int_{\xi_j-2d}^{\xi_j+2d} \frac{\partial \hat{h}_\varepsilon(x-\xi_j,\ P_j)}{\partial x} +$$

$$+ \int_{M_j-\sqrt{\varepsilon}}^{\xi_j-2d}\left(\frac{1}{2}\chi'(x-M_j)\left(\hat{h}_\varepsilon(x-\xi_{j-1},\ P_{j-1}) - \hat{h}_\varepsilon(x-\xi_j,\ P_j)\right) - \right.$$

$$\left. - \frac{\partial \hat{h}_\varepsilon(x-\xi_j,\ P_j)}{\partial x}\chi(x-M_j)\right)\chi(x-M_j)\,dx +$$

$$+ \int_{\xi_j+2d}^{M_{j+1}+\sqrt{\varepsilon}}\left(\frac{1}{2}\chi'(x-M_{j+1})\left(\hat{h}_\varepsilon(x-\xi_j,\ P_j) - \hat{h}_\varepsilon(x-\xi_{j+1},\ P_{j+1})\right) - \right.$$

$$\left. - \frac{\partial \hat{h}_\varepsilon(x-\xi_j,\ P_j)}{\partial x}\chi(x-M_{j+1})\right)(1-\chi(x-M_{j+1}))\,dx.$$

The first integral in the last expression is identically zero because of (2.2b). By (2.8a) and (2.3a) for $x \in [\xi_{j-1} + d,\ \xi_j - d]$ one has

$$|\hat{h}_\varepsilon(x - \xi_{j-1},\ P_{j-1}) - \hat{h}_\varepsilon(x - \xi_j,\ P_j)| \leq (P^* - P_*)\varepsilon^2 + O(\varepsilon^3). \tag{4.18}$$

Using this, (4.13) and (2.8b) one obtains

$$|(\psi_j(\mathbf{s}),\ \phi_{N+j}(\mathbf{s}))| \leq K_2\, \varepsilon^{3/2},$$

where constant $K_2 > 0$ does not depend on $\mathbf{s} \in \mathbb{B}_\varepsilon$, $j = 1, ..., N-1$ and ε.

c) Let us estimate inner products for 'neighbors' $(\psi_j(\mathbf{s}),\ \phi_{j-1}(\mathbf{s}))$ for $j \in \{1, ..., N-1\}$. By definitions (4.9) and (4.14) one has

$$\left|\frac{(\psi_j(\mathbf{s}),\ \phi_{j-1}(s))}{C_j(\mathbf{s})}\right| = \left|\int_{M_j-\sqrt{\varepsilon}}^{M_j+\sqrt{\varepsilon}}\left((1-\chi(x-M_j))\frac{\partial \hat{h}_\varepsilon(x-\xi_j,\ P_j)}{\partial P}\right)\chi(x-M_j)\,dx\right| \leq K_3\, \varepsilon^{3/2},$$

where we used again estimates (4.13), (4.17).

4.1 'Approximate Invariant' Manifold: Definition and Properties

d) Let us estimate $(\psi_{j+N}(\mathbf{s}),\ \phi_{j+N}(\mathbf{s}))$ for $j \in \{1,...,N-1\}$.

$$\frac{(\psi_{j+N},\ \phi_{j+N})}{C_{j+N}(\mathbf{s})} = -\int_{M_j+\sqrt{\varepsilon}}^{M_{j+1}-\sqrt{\varepsilon}} \frac{\partial \hat{h}_\varepsilon(x-\xi_j,\ P_j)}{\partial x} \int_{\xi_j}^x \frac{\hat{h}_\varepsilon(s-\xi_j,\ P_j) - \hat{h}_\varepsilon^-(P_j)}{\hat{h}_\varepsilon(s-\xi_j,\ P_j)^3}\,ds\,dx +$$

$$+ \int_{M_{j+1}-\sqrt{\varepsilon}}^{M_{j+1}+\sqrt{\varepsilon}} \left(\frac{1}{2}\chi'(x-M_{j+1})\left(\hat{h}_\varepsilon(x-\xi_j,\ P_j) - \hat{h}_\varepsilon(x-\xi_{j+1},\ P_{j+1})\right)\right.$$

$$\left. - \frac{\partial \hat{h}_\varepsilon(x-\xi_j,\ P_j)}{\partial x}(1-\chi(x-M_{j+1}))\right)\int_{\xi_j}^x \frac{\hat{h}_\varepsilon(s-\xi_j,\ P_j) - \hat{h}_\varepsilon^-(P_j)}{\hat{h}_\varepsilon(s-\xi_j,\ P_j)^3}\,ds\,dx +$$

$$+ \int_{M_j-\sqrt{\varepsilon}}^{M_j+\sqrt{\varepsilon}} \left(\frac{1}{2}\chi'(x-M_j)\left(\hat{h}_\varepsilon(x-\xi_{j-1},\ P_{j-1}) - \hat{h}_\varepsilon(x-\xi_j,\ P_j)\right)\right.$$

$$\left. - \frac{\partial \hat{h}_\varepsilon(x-\xi_j,\ P_j)}{\partial x}\chi(x-M_j)\right)\int_{\xi_j}^x \frac{\hat{h}_\varepsilon(s-\xi_j,\ P_j) - \hat{h}_\varepsilon^-(P_j)}{\hat{h}_\varepsilon(s-\xi_j,\ P_j)^3}\,ds\,dx. \quad (4.19)$$

Let us integrate by parts the first integral at the right-hand side of (4.19):

$$-\int_{M_j+\sqrt{\varepsilon}}^{M_{j+1}-\sqrt{\varepsilon}} \frac{\partial \hat{h}_\varepsilon(x-\xi_j,\ P_j)}{\partial x}\int_{\xi_j}^x \frac{\hat{h}_\varepsilon(s-\xi_j,\ P_j) - \hat{h}_\varepsilon^-(P_j)}{\hat{h}_\varepsilon(s-\xi_j,\ P_j)^3}\,ds\,dx = \frac{1}{C_{N+j}(\mathbf{s})} -$$

$$-\left(\hat{h}_\varepsilon(M_{j+1}-\sqrt{\varepsilon}-\xi_j,\ P_j) - \hat{h}_\varepsilon^-(P_j)\right)\int_{\xi_j}^{M_{j+1}-\sqrt{\varepsilon}} \frac{\hat{h}_\varepsilon(x-\xi_j,\ P_j) - \hat{h}_\varepsilon^-(P_j)}{\hat{h}_\varepsilon(x-\xi_j,\ P_j)^3}\,dx +$$

$$+\left(\hat{h}_\varepsilon(M_j+\sqrt{\varepsilon}-\xi_j,\ P_j) - \hat{h}_\varepsilon^-(P_j)\right)\int_{\xi_j}^{M_j-\sqrt{\varepsilon}} \frac{\hat{h}_\varepsilon(x-\xi_j,\ P_j) - \hat{h}_\varepsilon^-(P_j)}{\hat{h}_\varepsilon(x-\xi_j,\ P_j)^3}\,dx.$$

Using this and estimates (2.8a), (2.3a)–(2.3b) one obtains

$$\left|\int_{M_j+\sqrt{\varepsilon}}^{M_{j+1}-\sqrt{\varepsilon}} \frac{\partial \hat{h}_\varepsilon(x-\xi_j,\ P_j)}{\partial x}\int_{\xi_j}^x \frac{\hat{h}_\varepsilon(s-\xi_j,\ P_j) - \hat{h}_\varepsilon^-(P_j)}{\hat{h}_\varepsilon(s-\xi_j,\ P_j)^3}\,ds\,dx + 1/C_{N+j}(\mathbf{s})\right| \le \frac{K_4}{\varepsilon^3}\exp\left(-\frac{d}{\sqrt{2}\varepsilon}\right).$$

The rest terms in (4.19) one can estimate as in point **b)** using (4.13), (4.18) and (2.8b) by $O(\varepsilon^{3/2})$. Therefore, from (4.19) one ends up with

$$|(\psi_{j+N}(\mathbf{s}),\ \phi_{j+N}(\mathbf{s})) - 1| \le K_5\,\varepsilon^{3/2}.$$

e) The rest of inner products $(\psi_j(\mathbf{s}),\ \phi_j(\mathbf{s}))$, which were not considered yet, can be estimated in a similar way to one in points **b)-c)**. ∎

Proposition 4.4. *For every* $\mathbf{s} \in \mathbb{B}_\varepsilon$ *there exist functions*

$$\bar{\psi}_0(\mathbf{s}),\ \bar{\psi}_1(\mathbf{s}),...,\bar{\psi}_{2N-1}(\mathbf{s}) \in C^\infty(-L,\ L),$$

such that for all sufficiently small $\varepsilon > 0$ *and every* $j,\ k \in \{0,...2N-1\}$ *one has*

$$\left(\bar{\psi}_j(\mathbf{s}),\ \phi_k(\mathbf{s})\right) = \delta_{j,k}. \quad (4.20)$$

Moreover, there exists a positive number K *not depending on* $\mathbf{s},\varepsilon,j$ *such that*

$$\left\|\psi_j(\mathbf{s}) - \bar{\psi}_j(\mathbf{s})\right\|_{L^\infty(-L,L)} \le K\varepsilon^{3/2}. \quad (4.21)$$

Chapter 4 Formal Reduction onto an 'Approximate Invariant' Manifold

Proof: Let us search $\bar{\psi}_j(\mathbf{s})$ in the form

$$\bar{\psi}_j(\mathbf{s}) = \sum_{i=0}^{2N-1} B_i^j(\mathbf{s})\psi_i(\mathbf{s})$$

From (4.20) it necessarily follows that the vector $\begin{bmatrix} B_0^j, & B_1^j, & ..., & B_{2N-1}^j \end{bmatrix}^T$ is the solution to a linear system of $2N$ algebraic equations given as

$$\begin{pmatrix} (\psi_0, \phi_0) & (\psi_1, \phi_0) & \cdots & (\psi_{2N-1}, \phi_0) \\ \cdots & \cdots & \cdots & \cdots \\ (\psi_0, \phi_j) & (\psi_1, \phi_j) & \cdots & (\psi_{2N-1}, \phi_j) \\ \cdots & \cdots & \cdots & \cdots \\ (\psi_0, \phi_{2N-1}) & (\psi_1, \phi_{2N-1}) & \cdots & (\psi_{2N-1}, \phi_{2N-1}) \end{pmatrix} \begin{bmatrix} B_0^j \\ \cdots \\ B_j^j \\ \cdots \\ B_{2N-1}^j \end{bmatrix} = \begin{bmatrix} 0 \\ \cdots \\ 1 \\ \cdots \\ 0 \end{bmatrix}. \quad (4.22)$$

By Proposition 4.3 for sufficiently small $\varepsilon > 0$ matrix $A(\mathbf{s})$ of the system (4.22) has a diagonal dominance, and therefore invertible. Hence there exists a unique solution to (4.22) and existence of $\bar{\psi}_j(\mathbf{s})$ satisfying (4.20) is proved. Moreover, one has $A(\mathbf{s}) = \text{Id} + D(\mathbf{s})$, where by Proposition 4.3 $||D(\mathbf{s})||_2 \leq \text{const}\,\varepsilon^{3/2}$ uniformly in $\mathbf{s}$. Expanding the inverse to A_ε as a Neumann series

$$A^{-1} = \sum_{k=0}^{\infty} (-1)^k D^k,$$

one obtains from (4.22) that

$$|B_i^j(\mathbf{s}) - \delta_{i,j}| \leq \text{const}\,\varepsilon^{3/2}.$$

Estimate (4.21) follows from this and uniform bounds in $\mathbf{s} \in \mathbb{B}_\varepsilon$ by definition (4.14) for $||\psi_j(\mathbf{s})||_{L^\infty(-L,L)}, j = 0, ..., 2N-1$. ∎

Let us now for every $\mathbf{m} \in \mathbb{P}_\varepsilon$ define a linear operator $P_\mathbf{m}$ acting on $v \in L^\infty(-L, L)$:

$$P_\mathbf{m} v := \sum_{j=0}^{2N-1} \left(\bar{\psi}_j(\mathbf{s}), v\right) \phi_j(\mathbf{s}). \quad (4.23)$$

From the definition it is clear that the image of $P_\mathbf{m}$ belongs to $\mathbb{T}_\mathbf{m}\mathbb{P}_\varepsilon$ and from orthogonality conditions (4.20) it follows that $P_\mathbf{m}{}^2 = P_\mathbf{m}$. Thus, $P_\mathbf{m}$ are indeed projections on the tangent space $\mathbb{T}_\mathbf{m}\mathbb{P}_\varepsilon$. From definitions (4.9), (4.14) one can deduce that $||P_\mathbf{m}||_{\mathcal{L}(L^\infty(-L,L),\,L^\infty(-L,L))}$ is bounded uniformly in $\mathbf{m} \in \mathbb{P}_\varepsilon$, and $P_\mathbf{m}$ is Fréchet differentiable with respect to $\mathbf{m}$.

4.2 Decomposition in a Neighborhood of the Manifold

We start this section by showing that in a sufficiently small $L^\infty(-L, L)$ neighborhood of the 'approximate invariant' manifold $\mathbb{P}_\varepsilon$ every function $h(x)$ can be decomposed into the sum of some point $\mathbf{m} \in \mathbb{P}_\varepsilon$ and the remainder function v such that $P_\mathbf{m} v = 0$. Everywhere below $\mathcal{O}_{\delta_\varepsilon}(\mathbb{P}_\varepsilon)$ and $\mathcal{O}_{\delta_{1,\varepsilon}}(\partial\mathbb{P}_\varepsilon)$ denote L^∞ neighborhoods with diameters δ_ε and $\delta_{1,\varepsilon}$ of $\mathbb{P}_\varepsilon$ and its boundary $\partial\mathbb{P}_\varepsilon$, respectively.

Theorem 4.5. *There exists a positive constant ε_1, such that for all $\varepsilon \in (0, \varepsilon_1)$ there exist positive functions δ_ε, $\delta_{1,\varepsilon}$ and a nonlinear differentiable function $\pi_\varepsilon : \mathcal{O}_{\delta_\varepsilon}(\mathbb{P}_\varepsilon) \setminus \mathcal{O}_{\delta_{1,\varepsilon}}(\partial\mathbb{P}_\varepsilon) \to \mathbb{P}_\varepsilon$, which satisfies*

$$P_{\pi_\varepsilon(h)}(h - \pi_\varepsilon(h)) \equiv 0, \quad \text{for all } h \in \mathcal{O}_{\delta_\varepsilon}(\mathbb{P}_\varepsilon) \setminus \mathcal{O}_{\delta_{1,\varepsilon}}(\partial\mathbb{P}_\varepsilon).$$

4.2 Decomposition in a Neighborhood of the Manifold

Proof: Let us first show that the required projector can be constructed locally, for $\mathcal{O}_{\delta_\varepsilon}(\mathbf{m}_0)$, a neighborhood of each point $\mathbf{m}_0 \in \mathbb{P}_\varepsilon$. If $h \in \mathcal{O}_{\delta_\varepsilon}(\mathbf{m}_0)$ then $h = \mathbf{m}_0 + w$ with $\|w\|_{L^\infty} \leq \delta_\varepsilon$ and the required $\mathbf{m} = \pi_\varepsilon(h)$ should be find from equation

$$P_{\mathbf{m}}(\mathbf{m}_0 + w - \mathbf{m}) = 0. \qquad (4.24)$$

Recall that $\mathbb{P}_\varepsilon = \mathbf{m}_\varepsilon(\mathbb{B}_\varepsilon)$ where $\mathbf{m}_\varepsilon$ is a diffeomorphism between open set $\mathbb{B}_\varepsilon$ given by (4.4) and $\mathbb{P}_\varepsilon$. Therefore, there exist points $\mathbf{s}$, $\mathbf{s}_0 \in \mathbb{B}_\varepsilon$, such that $\mathbf{m}_0 = \mathbf{m}_\varepsilon(\mathbf{s}_0)$ and $\mathbf{m} = \mathbf{m}_\varepsilon(\mathbf{s})$. Moreover, using definition (4.23) one can rewrite (4.24) as:

$$\left(\mathbf{m}_\varepsilon(\mathbf{s}_0) + w - \mathbf{m}_\varepsilon(\mathbf{s}), \bar{\psi}_j(\mathbf{s})\right) = 0, \quad \text{for} \quad j = 0, ..., 2N-1. \qquad (4.25)$$

Define now a function $F_\varepsilon : \mathbb{R}^{2N} \times L^\infty(-L, L) \to \mathbb{R}^{2N}$ as

$$F_\varepsilon(\mathbf{s}, w)_j := \left(\mathbf{m}_\varepsilon(\mathbf{s}_0) + w - \mathbf{m}_\varepsilon(\mathbf{s}), \bar{\psi}_j(\mathbf{s})\right), \quad j = 0, ..., 2N-1.$$

Then one has $F_\varepsilon(\mathbf{s}_0, 0) = 0$ and

$$(\partial_{\mathbf{s}} F_\varepsilon(\mathbf{s}_0, 0)\delta\mathbf{s})_j = -(\mathbf{m}'_\varepsilon(\mathbf{s})\delta\mathbf{s}, \bar{\psi}_j(\mathbf{s}_0)) =$$
$$= -\sum_{i=0}^{i=2N-1}\left(\phi_i(\mathbf{s})\delta s_i, \bar{\psi}_j(\mathbf{s}_0)\right) = -\delta s_j,$$

where we denoted $\delta\mathbf{s} = [\delta s_0, \delta s_1, ..., \delta s_{2N-1}]$ and used orthogonality relations (4.20), which hold for sufficiently small $\varepsilon > 0$. From this it follows $D_{\mathbf{s}} F_\varepsilon(\mathbf{s}_0, 0) = -\text{Id}$, and therefore by the implicit function theorem there exist constant $\delta_\varepsilon > 0$ such that for all w with $\|w\|_{L^\infty} \leq \delta_\varepsilon$, equation $F_\varepsilon(\mathbf{s}, w) = 0$ has a unique solution $\mathbf{s} = \tilde{\mathbf{s}}_\varepsilon(w)$ or, what is the same, there exists unique $\mathbf{m} = \tilde{\mathbf{m}}_\varepsilon(w)$ satisfying equation (4.25). Moreover, there exists a positive function $\delta_{1,\varepsilon}$, such that one can choose δ_ε not depending on a choice of $\mathbf{m}_0 \in \mathbb{P}_\varepsilon \setminus \mathcal{O}_{\delta_{1,\varepsilon}}(\partial\mathbb{P}_\varepsilon)$. Therefore, one can construct the required projector π_ε globally on $\mathcal{O}_{\delta_\varepsilon}(\mathbb{P}_\varepsilon) \setminus \mathcal{O}_{\delta_{1,\varepsilon}}(\partial\mathbb{P}_\varepsilon)$. The differentiability of $\pi_\varepsilon(h)$ follows follows also from the implicit function theorem. Theorem is proved. ∎

In Bertozzi et al. [15] was shown that for every positive initial data $h_0 \in H^1(-L, L)$ with

$$\int_{-L}^{L} \frac{1}{2}|\partial_x h_0|^2 + U(h_0)\,dx < \infty,$$

for all $t > 0$ there exists a unique positive smooth solution $h(x,t)$ to (4.1) with boundary conditions (4.3) such that $h(x,0) = h_0(x)$. We restrict ourself to consider (4.1) in a small neighborhood $\mathcal{O}_{\delta_\varepsilon}(\mathbb{P}_\varepsilon) \setminus \mathcal{O}_{\delta_{1,\varepsilon}}(\partial\mathbb{P}_\varepsilon)$ of 'approximate invariant' manifold $\mathbb{P}_\varepsilon$. Taking δ_ε sufficiently small and using definition of $\mathbb{P}_\varepsilon$ one obtains that any $h_0 \in \mathcal{O}_{\delta_\varepsilon}(\mathbb{P}_\varepsilon) \cap H^1(-L, L)$ is positive, and therefore any solution to (4.1) with (4.3) such that $h(x,0) = h_0(x)$ exists for all $t > 0$. According to Theorem 4.5 such solution can be uniquely decomposed as follows:

$$h(t) = \mathbf{m}(t) + v(t), \quad \mathbf{m}(t) \in \mathbb{P}_\varepsilon \qquad (4.26a)$$
$$P_{\mathbf{m}(t)} v(t) \equiv 0. \qquad (4.26b)$$

Inserting this into equation (4.1) one can write it in an equivalent form

$$\partial_t v + \mathbb{F}_\varepsilon'(\mathbf{m}(t))v(t) = -\mathbb{F}_\varepsilon(\mathbf{m}(t)) - \mathbb{F}_\varepsilon(v(t), \mathbf{m}(t)) - \mathbf{m}'(t), \qquad (4.27)$$

where $\mathbb{F}_\varepsilon(v, \mathbf{m}) := \mathbb{F}_\varepsilon(v + \mathbf{m}) - \mathbb{F}_\varepsilon(\mathbf{m}) - \mathbb{F}_\varepsilon'(\mathbf{m})v$. Let us now differentiate (4.26b) with respect

Chapter 4 Formal Reduction onto an 'Approximate Invariant' Manifold

to time:
$$P'_{\mathbf{m}(t)}[\mathbf{m}'(t)]v(t) + P_{\mathbf{m}(t)}\,\partial_t v(t) \equiv 0.$$

Applying now projection $P_{\mathbf{m}(t)}$ to (4.27) and noting the last expression one gets a differential equation for $\mathbf{m}(t)$ on manifold $\mathbb{P}_\varepsilon$ in the following form:

$$(\mathrm{Id} - \mathbb{D}(\mathbf{m}(t))[\cdot]v(t))\mathbf{m}'(t) = P_{\mathbf{m}(t)}\left(-\mathbb{F}_\varepsilon(\mathbf{m}(t)) - \mathbb{F}_\varepsilon(v(t),\mathbf{m}(t))\right) - \mathbb{S}(\mathbf{m}(t))v(t), \qquad (4.28)$$

where

$$\mathbb{D}(\mathbf{m})[\delta\mathbf{m}]v := P'_{\mathbf{m}}[\delta\mathbf{m}]v,$$
$$\mathbb{S}(\mathbf{m})v := P_{\mathbf{m}}\left(\mathbb{F}_\varepsilon{}'(\mathbf{m})v\right). \qquad (4.29)$$

Let us also denote
$$\mathbb{M}(\mathbf{m},v)w := (\mathrm{Id} - \mathbb{D}(\mathbf{m})[\cdot]v)^{-1}w \qquad (4.30)$$

Then for each $v \in L^\infty(-L, L)$ such that $\|v\|_{L^\infty}$ is sufficiently small and each $\mathbf{m} \in \mathbb{P}_\varepsilon$ one has that operator $\mathbb{M}(\mathbf{m},v): L^\infty(-L, L) \to \mathbb{P}_\varepsilon$ is well defined and can be represented as Neumann series:

$$\mathbb{M}(\mathbf{m},v) = \sum_{i=0}^{\infty}(\mathbb{D}(\mathbf{m})[\cdot]v)^i.$$

Thus, equation (4.28) can be written in the following more convenient form:

$$\mathbf{m}'(t) = f(\mathbf{m}(t), v(t)), \qquad (4.31)$$

where
$$f(\mathbf{m},v) := \mathbb{M}(\mathbf{m},v)\left(-P_{\mathbf{m}}(\mathbb{F}_\varepsilon(\mathbf{m}) + \mathbb{F}_\varepsilon(v,\mathbf{m}))\right) - \mathbb{S}(\mathbf{m})v. \qquad (4.32)$$

We conclude that if $h(t)$ solves (4.1) and $h(t) \in \mathcal{O}_{\delta_\varepsilon}(\mathbb{P}_\varepsilon)\setminus\mathcal{O}_{\delta_{1,\varepsilon}}(\partial\mathbb{P}_\varepsilon)$ on $[0, T]$ then the associated functions $\mathbf{m}(t)$ and $v(t)$ satisfy on $[0, T]$ the following system:

$$\begin{cases} \partial_t v + \mathbb{F}_\varepsilon{}'(\mathbf{m}(t))v(t) = h(\mathbf{m}(t),\,v(t),\,\mathbf{m}'(t)) \\ \mathbf{m}'(t) = f(\mathbf{m},v) \end{cases}, \qquad (4.33)$$

where we denoted $h(\mathbf{m},v,w) := -\mathbb{F}_\varepsilon(\mathbf{m}) - \mathbb{F}_\varepsilon(v,\mathbf{m}) - w$. Vice versa, any solution $(\mathbf{m}(t), v(t))$ to (4.33) on $[0, T]$ with sufficiently small $v(t)$ satisfying $\mathbb{P}_{\mathbf{m}(0)}v(0) = 0$ generates a unique solution $u(t) := \mathbf{m}(t) + v(t)$ to equation (4.1). Therefore, instead of the initial lubrication equation (4.1) we can consider the associated system (4.33).

4.3 Formal Leading Order for Equation on the Manifold

In this section we formally assume that for any solution to (4.1) having at $t = 0$ an initial value in the neighborhood $\mathcal{O}_{\delta_\varepsilon}(\mathbb{P}_\varepsilon) \setminus \mathcal{O}_{\delta_{1,\varepsilon}}(\partial\mathbb{P}_\varepsilon)$ defined in Theorem 4.5, the norm of reminder $v(t)$ (obtained via decomposition (4.26a)) is negligible small for all $t > 0$ in comparison with that of $\mathbf{m}(t) \in \mathbb{P}_\varepsilon$. Then putting formally $v(t) \equiv 0$ for $t > 0$ we obtain a formal leading order

$$\mathbf{m}'(t) = f(\mathbf{m}(t), 0) \qquad (4.34)$$

for equation (4.31), which describes evolution of $\mathbf{m}(t)$ on manifold $\mathbb{P}_\varepsilon$. Below we transform (4.34) to an ODE system describing evolution of pressures $P_j(t)$ and positions $\xi_j(t)$ of multi-pulse structure $\mathbf{m}(t)$.

4.3 Formal Leading Order for Equation on the Manifold

Putting $v(t) \equiv 0$ into definitions (4.29)–(4.30), (4.32) one writes (4.34) as

$$\mathbf{m}'(t) = -P_\mathbf{m}\, \mathbb{F}_\varepsilon(\mathbf{m}).$$

Let us rewrite the last equation in a coordinate form on manifold $\mathbb{P}_\varepsilon$. Denoting as before

$$\mathbf{s} = (s_0, ..., s_{2N-1}) := (P_0, ..., P_N, \xi_1, ..., \xi_{N-1})$$

such that $\mathbf{m} = \mathbf{m}_\varepsilon(\mathbf{s})$ and taking the standard scalar product in $L^2(-L, L)$ of $\mathbf{m}'(t)$ with $\bar{\psi}_j(\mathbf{s})$ for $j = 0, ..., 2N - 1$ one gets

$$(\mathbf{m}'(t), \bar{\psi}_j(\mathbf{s})) = \sum_{i=0}^{i=2N-1} \left(\phi_i(\mathbf{s}) \frac{ds_i}{dt}, \bar{\psi}_j(\mathbf{s})\right) = \frac{ds_j}{dt}, \qquad (4.35)$$

where we used definition (4.9) and orthogonality conditions (4.20). On the other hand

$$(P_\mathbf{m}\, \mathbb{F}_\varepsilon(\mathbf{m}_\varepsilon(\mathbf{s})), \bar{\psi}_j(\mathbf{s})) = (\mathbb{F}_\varepsilon(\mathbf{m}_\varepsilon(\mathbf{s})), \bar{\psi}_j(\mathbf{s}))$$

By Proposition 4.4 for all $\mathbf{s} \in \mathbb{B}_\varepsilon$ and $j = 0, ..., 2N - 1$ one has

$$(\mathbb{F}_\varepsilon(\mathbf{m}_\varepsilon(\mathbf{s})), \bar{\psi}_j(\mathbf{s})) \sim (\mathbb{F}_\varepsilon(\mathbf{m}_\varepsilon(\mathbf{s})), \psi_j(\mathbf{s})) \text{ as } \varepsilon \to 0. \qquad (4.36)$$

Next, by definition (4.14) and representation (4.11) one has for $j = 1, ..., N - 1$

$$(\mathbb{F}_\varepsilon(\mathbf{m}_\varepsilon(\mathbf{s})), \psi_j(\mathbf{s})) = C_j(\mathbf{s}) \Big(\int_{M_j-\sqrt{\varepsilon}}^{M_j+\sqrt{\varepsilon}} \chi(x - M_j) \mathbb{F}_\varepsilon(\mathbf{m})(x)\, dx +$$

$$+ \int_{M_{j+1}-\sqrt{\varepsilon}}^{M_{j+1}+\sqrt{\varepsilon}} (1 - \chi(x - M_{j+1})) \mathbb{F}_\varepsilon(\mathbf{m})(x)\, dx \Big) =$$

$$= C_j(\mathbf{s}) \left(\int_{\theta_j}^{M_j+\sqrt{\varepsilon}} \mathbb{F}_\varepsilon(\mathbf{m})(x)\, dx + \int_{M_{j+1}-\sqrt{\varepsilon}}^{\theta_{j+1}} \mathbb{F}_\varepsilon(\mathbf{m})(x)\, dx \right)$$

$$= \frac{J(\mathbf{s})(\theta_{j+1}) - J(\mathbf{s})(\theta_j)}{\int_{M_j+\sqrt{\varepsilon}}^{M_{j+1}-\sqrt{\varepsilon}} \frac{\partial \bar{h}_\varepsilon(x-\xi_j,\, P_j)}{\partial P}\, dx},$$

where we used the second mean-value theorem for integration (see paragraph 1.13 in Jeffreys and Jeffreys [45]) with θ_j being some point in $(M_j - \sqrt{\varepsilon}, M_j + \sqrt{\varepsilon})$ and introduced for each $\mathbf{s} \in \mathbb{B}_\varepsilon$ a flux function $J(\mathbf{s}) \in C^\infty(-L, L)$ as

$$J(\mathbf{s}) := \mathbf{m}_\varepsilon(\mathbf{s})^3 \partial_x \left(-\Pi_\varepsilon(\mathbf{m}_\varepsilon(\mathbf{s})) + \partial_{xx} \mathbf{m}_\varepsilon(\mathbf{s})\right). \qquad (4.37)$$

Chapter 4 Formal Reduction onto an 'Approximate Invariant' Manifold

Analogously, by definition (4.14) for $j = 1, ..., N-1$:

$$(\mathbb{F}_\varepsilon(\mathbf{m}_\varepsilon(\mathbf{s})), \psi_{j+N}(\mathbf{s})) =$$

$$= C_{N+j}(\mathbf{s}) \Big(\int_{M_j-\sqrt{\varepsilon}}^{M_j+\sqrt{\varepsilon}} \chi(x - M_j) \int_{\xi_j}^x \frac{\hat{h}_\varepsilon(s - \xi_j, P_j) - \hat{h}_\varepsilon^-(P_j)}{\hat{h}_\varepsilon(s - \xi_j, P_j)^3} ds \, \mathbb{F}_\varepsilon(\mathbf{m})(x) \, dx +$$

$$+ \int_{M_{j+1}-\sqrt{\varepsilon}}^{M_{j+1}+\sqrt{\varepsilon}} (1 - \chi(x - M_{j+1})) \int_{\xi_j}^x \frac{\hat{h}_\varepsilon(s - \xi_j, P_j) - \hat{h}_\varepsilon^-(P_j)}{\hat{h}_\varepsilon(s - \xi_j, P_j)^3} ds \, \mathbb{F}_\varepsilon(\mathbf{m})(x) \, dx \Big) =$$

$$= C_{N+j}(\mathbf{s}) \Big(\int_{\xi_j}^{M_j+\sqrt{\varepsilon}} \frac{\hat{h}_\varepsilon(x - \xi_j, P_j) - \hat{h}_\varepsilon^-(P_j)}{\hat{h}_\varepsilon(x - \xi_j, P_j)^3} dx \int_{\theta_j}^{M_j+\sqrt{\varepsilon}} \mathbb{F}_\varepsilon(\mathbf{m})(x) \, dx +$$

$$+ \int_{\xi_j}^{M_{j+1}-\sqrt{\varepsilon}} \frac{\hat{h}_\varepsilon(x - \xi_j, P_j) - \hat{h}_\varepsilon^-(P_j)}{\hat{h}_\varepsilon(x - \xi_j, P_j)^3} dx \int_{M_{j+1}-\sqrt{\varepsilon}}^{\theta_{j+1}} \mathbb{F}_\varepsilon(\mathbf{m})(x) \Big) dx =$$

$$= \frac{\int_{M_j+\sqrt{\varepsilon}}^{M_{j+1}-\sqrt{\varepsilon}} \frac{\hat{h}_\varepsilon(x-\xi_j, P_j)-\hat{h}_\varepsilon^-(P_j)}{\hat{h}_\varepsilon(x-\xi_j, P_j)^3} dx}{2 \int_{M_j+\sqrt{\varepsilon}}^{M_{j+1}-\sqrt{\varepsilon}} \frac{(\hat{h}_\varepsilon(x-\xi_j, P_j)-\hat{h}_\varepsilon^-(P_j))^2}{\hat{h}_\varepsilon(x-\xi_j, P_j)^3} dx} (J(\mathbf{s})(\theta_{j+1}) + J(\mathbf{s})(\theta_j)). \quad (4.38)$$

Finally, denoting

$$J_{j-1,j} := J(\mathbf{s})(\theta_j), \; j = 1, ..., N-1$$
$$J_{-1,0} := -J_{0,1}, \quad J_{N,N+1} := -J_{N-1,N} \quad (4.39)$$

and combining (4.35), (4.36), (4.38) together one obtains the following coordinate form for the leading order of equation (4.34) as $\varepsilon \to 0$:

$$\frac{dP_j}{dt} = C_{P,j} \cdot (J_{j,j+1} - J_{j-1,j}),$$
$$\frac{d\xi_j}{dt} = -C_{\xi,j} \cdot (J_{j,j+1} + J_{j-1,j}), \; j = 0, ..., N \quad (4.40)$$

where we denoted for $j = 1, ..., N-1$:

$$C_{P,j} := -1 \Big/ \int_{M_j+\sqrt{\varepsilon}}^{M_{j+1}-\sqrt{\varepsilon}} \frac{\partial \hat{h}_\varepsilon(x - \xi_j, P_j)}{\partial P} dx,$$

$$C_{\xi,j} := \frac{\int_{M_j+\sqrt{\varepsilon}}^{M_{j+1}-\sqrt{\varepsilon}} \frac{h_\varepsilon(x-\xi_j, P_j)-\hat{h}_\varepsilon^-(P_j)}{\hat{h}_\varepsilon(x-\xi_j, P_j)^3} dx}{2 \int_{M_j+\sqrt{\varepsilon}}^{M_{j+1}-\sqrt{\varepsilon}} \frac{(\hat{h}_\varepsilon(x-\xi_j, P_j)-\hat{h}_\varepsilon^-(P_j))^2}{\hat{h}_\varepsilon(x-\xi_j, P_j)^3} dx};$$

and

$$C_{P,0} := -1 \Big/ \Big(2 \int_{-L}^{M_1-\sqrt{\varepsilon}} \frac{\partial h_0}{\partial P} dx \Big),$$
$$C_{P,N} := -1 \Big/ \Big(2 \int_{M_N+\sqrt{\varepsilon}}^{L} \frac{\partial h_N}{\partial P} dx \Big). \quad (4.41)$$

We conclude that (4.40)–(4.41) gives us a new reduced ODE model describing evolution of pressures and positions in multi-pulse structures (droplet arrays) governed by the no-slip lubrication model (1.6) In the next section we discuss a correspondence between the new reduced ODE

(4.40) model and previously asymptotically derived (2.66) corresponding to the no-slip lubrication equation.

4.4 Comparison of Reduced ODE Models

Comparing (4.40)–(4.41) with ODE reduced model (2.66) considered with coefficients (2.27)–(2.28) and mobility term $M(h) = h^3$ one can see that formally they have the same form. Differences between them sit in the formulas for coefficients $C_{P,j}$, $C_{\xi,j}$ and fluxes $J_{j-1,j}$. Below we state these differences and point out some advantages of system (4.40) in comparison with asymptotically derived one (2.66).

- In definitions (4.41) interval of integration $[M_j+\sqrt{\varepsilon}, M_{j+1}-\sqrt{\varepsilon}]$ corresponds to one $[-\tilde{L}, \tilde{L}]$ from section 2.3.1, the support of the j-th droplet in an array of $N+1$ ones. But in contrast to the latter one given positions ξ_j, $j = 1, ..., N-1$ of droplets in the array the interval $[M_j+\sqrt{\varepsilon}, M_{j+1}-\sqrt{\varepsilon}]$ can be calculated explicitly using formula (4.5). In the case of system (2.66) it is not clear how to estimate the droplet support $[-\tilde{L}, \tilde{L}]$.

- Fluxes $J_{j-1,j}$ between neighboring droplets in system (4.40) are defined using explicit formula (4.37), (4.39). In the case of system (2.66) fluxes $J_{j-1,j}$ are not defined explicitely and we can use only asymptotic approximations for them given by (2.56). Although points $\theta_j \in (M_j - \sqrt{\varepsilon}, M_j + \sqrt{\varepsilon})$ in definition (4.39) arise after the application of a mean-value type theorem and are not given explicitly, nevertheless definitions (4.37), (4.39) are more suitable for a rigorous analysis of the reduced ODE models.

- Both reduced ODE models give us in some sense a leading order approximation for the evolution of solutions to the no-slip equation, but derivation of system (4.40) based on the 'approximate invariant' manifold approach gives new possibilities for analytical error estimates of such approximations. Namely, using Propositions 4.1, 4.4 and formula (4.36) one can estimate that passing from equation (4.34) to system (4.40) one skips terms which are $O(\varepsilon^3)$. Moreover, having a rigorous estimate on the smallness of remainder function $v(t)$ from decomposition (4.26a) one could similarly estimate the smallness of terms, which one skips during passing from the exact equation (4.31) on the 'approximate invariant' manifold $\mathbb{P}_\varepsilon$ to its leading order (4.34).

4.5 Discussion and Spectral Problem

The above derivation of the reduced ODE model (4.40) is inspired by a recent article of Mielke and Zelik [4] which proves a center manifold reduction theorem for a general class of semilinear parabolic equations which posses multi-pulse solutions. Proceeding similar to it and writing formally the leading order for the reduced system corresponding to (4.1) with (4.3) we do not prove here a center-manifold existence theorem. The reason for that are additional difficulties for applying the approach of Mielke and Zelik [4] to the no-slip lubrication model. The most important are:

- The operator (4.2) is not semi-linear but rather quasilinear and what is more difficult to handle is, that it degenerates as $h \to 0$.

- Preliminary asymptotical (see Glasner [39]) and our numerical analysis in paragraph 2.6.2 showed that the reduced ODE models are valid in the limit $\varepsilon \to 0$. In this case one of the main assumption for proving a type of a center-manifold existence theorem is one concerning the spectrum of operator (4.2) linearized at a point $\mathbf{m} \in \mathbb{P}_\varepsilon$ as $\varepsilon \to 0$. In

Chapter 4 Formal Reduction onto an 'Approximate Invariant' Manifold

the next chapter we derive rigorously the spectrum asymptotics for (4.2) linearized at the stationary solution $h_{0,\varepsilon}$, which describes physically a droplet on a bounded interval. It turns out that the corresponding linearized eigenvalue problem is a singular perturbed one and the spectrum of it tends to zero as $\varepsilon \to 0$. Such behavior of the spectrum of the linearized problem is quite different from the main spectrum assumption used in the approach of Mielke and Zelik [4] (see assumptions (2.20) and (2.25) there).

- Moreover, we expect due to observations above that the construction of a center-manifold will be singular-perturbed problem as well. Up to our knowledge there are no many results on the rigorous existence of singularly perturbed center-manifolds, except for Fenichel theory for ODEs (see Fenichel [46] and Jones [47]) and several partial examples for PDEs.

Nevertheless, the good correspondence between reduced models (4.40) and (2.66) indicates an applicability of a center-manifold reduction approach for the rigorous justification of them in future. An important analytical result pointing out such a possibility is derived in the next chapter and states existence of ε dependent gap in the spectrum of the no-slip lubrication equation linearized at the stationary solution $h_{0,\varepsilon}$ (see Remark 5.15).

Chapter 5
Spectrum Asymptotics in a Singular Limit

In this chapter we describe linear stability properties for the no-slip equation (1.6) with boundary conditions (1.10) at a stationary solution which corresponds physically to a single droplet on a bounded interval. It turns out that the underlying linearized eigenvalue problem (EVP) is a singularly perturbed one. Up to our knowledge here for the first time rigorous results by means of a singularly perturbed analysis for the spectrum of a linearized thin film type equation such as (1.6) in the limit $\varepsilon \to 0$, where the parameter ε appears in the pressure function (1.4), are derived. The main result of this chapter–existence of an ε dependent spectral gap is derived using a combined application of three analytical approaches. The first one establishes the rigorous asymptotics for the droplet stationary solution above as $\varepsilon \to 0$ and is summarized in Lemmata 5.8 and 5.23. The second approach concerns with approximate eigenvalue problems. Asymptotics for the spectra of these problems as $\varepsilon \to 0$ is investigated here using construction and analysis of corresponding characteristic determinants (in a way similar to that described by Kamke [48], second part, paragraph I.2.1). A typical application of this approach is the proof of Theorem 5.24. The third approach is based on applications of a modified implicit function theorem first introduced by Magnus [41] and Recke and Omel'chenko [5] to the proof of existence of eigenvalues with prescribed asymptotics as $\varepsilon \to 0$, in particular of a unique exponentially small one. Typical applications of this approach are proofs of Theorems 5.12 and 5.29. We hope also that the last approach can be used in future for showing existence of solutions to a certain class of singularly perturbed eigenvalue problems.

5.1 Scalings and Linearized Eigenvalue Problems

In this chapter we introduce a scaling for the no-slip lubrication model, and therefore use a slightly different notation for its variables in comparison with previous chapters. Let us write the no-slip lubrication model defined in (1.6) in the form:

$$\partial_{\bar{t}} \bar{h} = -\partial_{\bar{x}} \left(\bar{h}^3 \partial_{\bar{x}} \left(\partial_{\bar{x}\bar{x}} \bar{h} - \overline{\Pi}\left(\bar{h}\right) \right) \right), \tag{5.1}$$

with boundary conditions on a fixed interval $[-L, L]$

$$\partial_{\bar{x}\bar{x}\bar{x}} \bar{h} = 0, \quad \text{and} \quad \partial_{\bar{x}} \bar{h} = 0 \quad \text{at} \quad \bar{x} = \pm L, \tag{5.2}$$

which incorporate zero flux at the boundary and imply conservation of mass law

$$\bar{h}_c \equiv \frac{1}{2L} \int_{-L}^{+L} \bar{h}(\bar{x}, \bar{t}) \, d\bar{x} \quad \text{for all } \bar{t} > 0,$$

and the potential function

$$\overline{\Pi}_\varepsilon \left(\bar{h} \right) = \frac{\varepsilon^2}{\bar{h}^3} - \frac{\varepsilon^3}{\bar{h}^4}.$$

Chapter 5 Spectrum Asymptotics in a Singular Limit

We call (5.1) with (5.2) a unscaled (physically relevant) version of the no-slip lubrication equation. Note that in this chapter (in contrast to previous ones) the unscaled variables and functions are denoted with a overline sign. Next theorem summarizes results from Bertozzi et al. [15], Glasner and Witelski [29] on stationary solutions to (5.1) considered on a bounded interval with boundary conditions (5.2).

Theorem 5.1. *Equation (5.1) with (5.2) has a family of positive nonconstant steady state solutions $\overline{h}_{0,\varepsilon}(\overline{x}, P)$ parameterized by a constant (a so called pressure) $P \in (0, P_{max}(\varepsilon))$, where $P_{max}(\varepsilon)$ is defined in (2.1), which satisfy*

$$\partial_{\overline{x}\overline{x}}\overline{h}_{0,\varepsilon}(\overline{x}, P) = \overline{\Pi}_\varepsilon\left(\overline{h}_{0,\varepsilon}(\overline{x}, P)\right) - P, \tag{5.3a}$$

$$\overline{h}_{0,\varepsilon}(\overline{x}, P) = \overline{h}_{0,\varepsilon}(-\overline{x}, P), \tag{5.3b}$$

$$\partial_{\overline{x}}\overline{h}_{0,\varepsilon}(0, P) = 0 \quad and \quad \partial_{\overline{x}}\overline{h}_{0,\varepsilon}(\overline{x}, P) < 0 \ for \ \overline{x} \in (0, L). \tag{5.3c}$$

Proof: It is simply to deduce that a solution to (5.1) with (5.2) is stationary if and only if it satisfies (5.3a) with (5.2). The rest of the proof can be done via a phase plane analysis as in the proof of Theorem 2.1. It shows that for a fixed $P \in (0, P_{max}(\varepsilon))$ there exists a family of periodic orbits to the equation $\overline{h}''(\overline{x}) = \Pi_\varepsilon\left(\overline{h}(\overline{x})\right) - P$ nested into a homoclinic loop. The family can be parameterized by the least period $T = 2L$. For any orbit there exists a phase shift such that the corresponding periodic solution restricted to the interval $[-L, L]$ gives a smooth stationary solution $\overline{h}_{0,\varepsilon}(\overline{x}, P)$ to (5.1)–(5.2) satisfying (5.3b)–(5.3c). ∎

In this chapter we fix P and L so that

$$L - A/P > 0 \text{ with } A := \frac{1}{\sqrt{3}}. \tag{5.4}$$

Assumption (5.4) allows us below to distinguish three different asymptotic regions for the stationary solution as $\varepsilon \to 0$ (see Lemma 5.8 and Remark 5.9) and is important for all results of this chapter. Define next a linear operator $\overline{\mathcal{L}}_\varepsilon$ acting in $L^2(-L, L)$ and induced by the linearization of (5.1)–(5.2) at the steady state $\overline{h}_{0,\varepsilon}(\overline{x}, P)$,

$$\overline{\mathcal{L}}_\varepsilon = -\frac{d}{d\overline{x}}\left[\overline{h}_{0,\varepsilon}^3 \frac{d}{d\overline{x}}\left(\frac{d^2}{d\overline{x}^2} \cdot - \overline{\Pi}'_\varepsilon(\overline{h}_{0,\varepsilon}) \cdot\right)\right],$$

where

$$D\left(\overline{\mathcal{L}}_\varepsilon\right) := \left\{\overline{\eta} \in W^{4,2}(-L, L) : \ \overline{\eta}'''(\pm L) = \overline{\eta}'(\pm L) = \int_{-L}^{L} \overline{\eta} \, d\overline{x} = 0\right\}.$$

The integral constraint in the last definition is induced by conservation of mass (i.e. we are practically interested only in such perturbations $\overline{\eta}(\overline{x})$ of the steady state $\overline{h}_{0,\varepsilon}(\overline{x}, P)$ that preserve its mass). The eigenvalue problem associated with $\overline{\mathcal{L}}_\varepsilon$ we write as

$$\overline{\mathcal{L}}_\varepsilon \overline{\eta} = -\overline{\lambda}\overline{\eta}, \ \overline{\eta} \in D\left(\overline{\mathcal{L}}_\varepsilon\right). \tag{5.5}$$

Let us next introduce scalings

$$x = \frac{\overline{x}}{\varepsilon}, \quad h = \frac{\overline{h}}{\epsilon}, \quad t = \frac{\overline{t}}{\varepsilon}, \tag{5.6}$$

and apply them to the variables of (5.1)–(5.2). These scalings were first used for (5.1) in section

5.2 of Glasner [39]. In the new variables the no-slip lubrication equation has a form

$$\partial_t h = -\partial_x \left(h^3 \partial_x \left(\partial_{xx} h - \Pi(h) \right) \right), \tag{5.7}$$

where

$$\Pi(h) := \varepsilon \overline{\Pi}_\varepsilon(h\,\varepsilon) = \frac{1}{h^3} - \frac{1}{h^4}. \tag{5.8}$$

On solution of (5.7) the following boundary conditions are imposed:

$$\partial_{xxx} h = 0, \quad \text{and} \quad \partial_x h = 0 \quad \text{at} \quad x = \pm L/\varepsilon, \tag{5.9}$$

which incorporate zero flux at the boundary and imply again the conservation of mass law

$$h_c \equiv \frac{\varepsilon}{2L} \int_{-L/\varepsilon}^{L/\varepsilon} h(x,t)\,dx, \quad \forall t > 0.$$

Let us define a function

$$h_{0,\varepsilon}(x) := \frac{\overline{h}_{0,\varepsilon}(\varepsilon x, P)}{\varepsilon}. \tag{5.10}$$

Proposition 5.2. *For each $\varepsilon > 0$ function $h_{0,\varepsilon}(x)$ is a stationary solution to (5.7) with (5.9) and satisfies*

$$h_{0,\varepsilon}''(x) = \Pi\left(h_{0,\varepsilon}(x)\right) - \varepsilon\,P, \tag{5.11a}$$
$$h_{0,\varepsilon}(x) = h_{0,\varepsilon}(-x), \tag{5.11b}$$
$$h_{0,\varepsilon}'(0) = 0 \quad \text{and} \quad h_{0,\varepsilon}'(x) < 0 \text{ for } x \in (0, L/\varepsilon). \tag{5.11c}$$

Proof: The validity of (5.11a)–(5.11c) one checks by the direct substitution of definition (5.10) for $h_{0,\varepsilon}(x)$ into above expressions and using (5.3a)–(5.3c). Similarly using that $\overline{h}_{0,\varepsilon}(\overline{x}, P)$ satisfies boundary conditions (5.2) one shows that $h_{0,\varepsilon}(x)$ satisfies (5.9). Finally, any solution to (5.11a) satisfies also (5.7). ■

Remark 5.3. Note that the parameter pressure P is fixed throughout the whole chapter and all asymptotics are considered with respect to the parameter $\varepsilon \to 0$. This explains the choice of the scaling (5.10) for the stationary solution and an appearance of the term εP in (5.11a). ■

The corresponding to (5.7) right-hand side operator linearized at the scaled stationary solution $h_{0,\varepsilon}$ is given by

$$\mathcal{L}_\varepsilon = -\frac{d}{dx}\left[h_{0,\varepsilon}^3 \frac{d}{dx}\left(\frac{d^2}{dx^2}\cdot - \Pi'(h_{0,\varepsilon})\cdot\right)\right],$$

where

$$D\left(\mathcal{L}_\varepsilon\right) = \left\{\eta \in W^{4,2}(-L/\varepsilon, L/\varepsilon):\ \eta'''(\pm L/\varepsilon) = \eta'(\pm L/\varepsilon) = \int_{-L/\varepsilon}^{L/\varepsilon}\eta\,dx = 0\right\}.$$

The scaled EVP associated with operator $\mathcal{L}_\varepsilon$ one writes again as

$$\mathcal{L}_\varepsilon \eta = -\lambda \eta,\ \eta \in D\left(\mathcal{L}_\varepsilon\right). \tag{5.12}$$

One can easily derive the following relation between solutions of (5.5) and its scaled version (5.12):

Chapter 5 Spectrum Asymptotics in a Singular Limit

- the relation between eigenfunctions of (5.5) and (5.12):

$$\eta(x) := \frac{\overline{\eta}(\varepsilon x)}{\varepsilon};$$

- the relation between eigenvalues of (5.5) and (5.12):

$$-\lambda = -\overline{\lambda}\varepsilon. \tag{5.13}$$

For a fixed $\varepsilon > 0$ operator $\mathcal{L}_\varepsilon$ is a particular case of a general class of linear operators associated with linearized thin film type equations. For such operators qualitative properties of their spectra were investigated by Laugesen and Pugh [49]. For our subsequent purposes we summarize them here applied to $\mathcal{L}_\varepsilon$. Firstly, we use a transformation of the EVP (5.12) to a symmetric one. Define functions

$$r_\varepsilon(x) := -\Pi'(h_{0,\varepsilon}(x)), \tag{5.14a}$$
$$f_\varepsilon(x) := (h_{0,\varepsilon}(x))^{-3}. \tag{5.14b}$$

In Appendix B of Laugesen and Pugh [49] was shown that if a pair $[\eta, \lambda]$ is a solution to the initial EVP (5.12), then a pair $[h, \lambda]$ with

$$h(x) := \int_{-L/\varepsilon}^{x} \eta(s)\, ds$$

satisfies

$$h^{(4)}(x) + (r_\varepsilon(x) h'(x))' = \lambda f_\varepsilon(x) h(x), \tag{5.15a}$$
$$h''(\pm L/\varepsilon) = h(\pm L/\varepsilon) = 0. \tag{5.15b}$$

Vice versa any solution $[h, \lambda]$ to (5.15a)–(5.15b) gives a solution $[\eta, \lambda]$ to (5.12) with $\eta(x) := h'(x)$.

We define next Hilbert spaces

$$W_\varepsilon := H^2(-L/\varepsilon, L/\varepsilon) \cap H^1_0(-L/\varepsilon, L/\varepsilon) \tag{5.16}$$

equipped with the standard $H^2(-L/\varepsilon, L/\varepsilon)$ inner product and $L^2(-L/\varepsilon, L/\varepsilon)$ with a weighted one:

$$(h, w)_\varepsilon := \int_{-L/\varepsilon}^{L/\varepsilon} h\, w\, f_\varepsilon\, dx. \tag{5.17}$$

The next theorem is a reformulation of Theorem 23 of Laugesen and Pugh [49] for our case.

Theorem 5.4. *Consider a symmetric EVP*

$$h \in W_\varepsilon,\ \lambda \in \mathbb{R}\ :\ \int_{-L/\varepsilon}^{L/\varepsilon} (h'' w'' - r_\varepsilon h' w' - \lambda f_\varepsilon h\, w)\, dx = 0,\ \forall w \in W_\varepsilon. \tag{5.18}$$

For a fixed $\varepsilon > 0$ there exist sequences $\{\lambda_\varepsilon^*, \lambda_\varepsilon^0, \lambda_\varepsilon^1, \lambda_\varepsilon^2, ...\}$, $\{h_\varepsilon^*, h_\varepsilon^0, h_\varepsilon^1, h_\varepsilon^2, ...\}$ such that:

(i) *for each $j \in \{*, 0, 1, 2, ...\}$ the pair $[h_\varepsilon^j, \lambda_\varepsilon^j]$ is a solution to (5.18);*

(ii)

$$\lambda_\varepsilon^* \leq \lambda_\varepsilon^0 \leq \lambda_\varepsilon^1 \leq \lambda_\varepsilon^2 \leq ... \to \infty;$$

(iii) the set of eigenfunctions h_ε^j, $j \in \{*, 0, 1, 2, ...\}$ forms an orthonormal basis in $L^2(-L/\varepsilon, L/\varepsilon)$ with respect to the weighted inner product (5.17). Moreover h_ε^j are C^4 smooth on $[-L/\varepsilon, L/\varepsilon]$ and the corresponding pair $[h_\varepsilon^j, \lambda_\varepsilon^j]$ satisfies (5.15a)–(5.15b).

Remark 5.5. As in Laugesen and Pugh [49] the transformation procedure stated above and the last theorem make it natural for us to investigate equivalent symmetric EVP problem (5.18) instead of the initial one (5.12). For the unscaled thin film equation (5.1)–(5.2) it is known that it is not uniformly elliptic as $h \to 0$ and degenerates in this limit. As a consequence the scaled system (5.7)–(5.9) and corresponding EVPs (5.12) and (5.18) have a singularity at $\varepsilon = 0$. When $\varepsilon = 0$ the assertions of the Proposition 5.2 are not valid anymore and one can not define a linearization of (5.7)–(5.9) at $h_{0,\varepsilon}$, because the latter even does not exists. This implies that the EVPs (5.12) and (5.18) are essentially singularly perturbed ones. ■

Remark 5.6. Finally, an application of Theorem 4 of Laugesen and Pugh [49] and Proposition 5.2 above to EVP (5.12) states that for any $\varepsilon > 0$ its largest eigenvalue is equal to $-\lambda_\varepsilon^*$ (where λ_ε^* is defined in Theorem 5.4) and positive. Using relation (5.13) between eigenvalues of EVPs (5.12) and (5.5) one concludes that the unscaled (physically relevant) equation (5.1) with (5.2) is linear unstable at the stationary solution $\overline{h}_{0,\varepsilon}$. Nevertheless, we show in this chapter that for sufficiently small $\varepsilon > 0$ EVP (5.5) has exactly one positive eigenvalue. ■

Remark 5.7. In Lemma 5.17 we derive asymptotics for functions (5.14a) and (5.14b) as $\varepsilon \to 0$, using asymptotics for $h_{0,\varepsilon}(x)$ stated in Lemma 5.8. In particularly, there we show that $f_\varepsilon(x)$ is positive, bounded and bounded from below away from zero, and that $r_\varepsilon(x)$ is bounded, as functions of x uniformly in $\varepsilon > 0$. Here we prefer to work with the scaled version of the no-slip equation (5.7) with (5.9) and corresponding EVPs because of an advantage of L_∞ bounds, holding uniformly in ε, for the coefficients of the symmetric eigenvalue problem (5.18). This uniformity is explored often in the proofs below. We should also point out that the no-slip lubrication model in the form (5.7) with pressure function (5.8) was already considered by Glasner et al. [3]. ■

5.2 Summary of Main Results and Discussion

First we formulate our main result concerning rigorous asymptotics for stationary solution $h_{0,\varepsilon}(x)$ as $\varepsilon \to 0$.

Lemma 5.8. *There exist $\tilde{\varepsilon} > 0$ and functions $a_\varepsilon, b_\varepsilon : (0, \tilde{\varepsilon}) \to \mathbb{R}$ such that the following assertions hold:*

(i) for all $\varepsilon \in (0, \tilde{\varepsilon})$ one has $0 < a_\varepsilon < b_\varepsilon < L$;

(ii) $a_\varepsilon, b_\varepsilon \to A/P$ and $b_\varepsilon - a_\varepsilon = O(\varepsilon^{1/4}) \to 0$ as $\varepsilon \to 0$, where A is defined in (5.4);

(iii) for all $x \in [0, a_\varepsilon/\varepsilon]$ it holds: $\varepsilon^{-3/4} \leq h_{0,\varepsilon}(x) = O(1/\varepsilon)$ and

$$h_{0,\varepsilon}(x) \sim \frac{P\varepsilon}{2}\left(\left(\frac{A}{P\varepsilon}\right)^2 - x^2\right);$$

(iv) for all $x \in [b_\varepsilon/\varepsilon, L/\varepsilon]$ it holds: $1 \leq h_{0,\varepsilon}(x) = 1 + O(\varepsilon^{1/6})$.

The proof of this lemma is given in section 5.4.

Chapter 5 Spectrum Asymptotics in a Singular Limit

Remark 5.9. In this chapter we call three intervals $(0, \ a_\varepsilon/\varepsilon)$, $(a_\varepsilon/\varepsilon, \ b_\varepsilon/\varepsilon)$, $(b_\varepsilon/\varepsilon, \ L/\varepsilon)$ as **droplet core**, **contact line** and **outer layer**, respectively. This notation formally corresponds to the physical relevant notions used by Glasner and Witelski [2] and is justified by the fact that on such defined droplet core and outer layer stationary solution $h_{0,\varepsilon}(x)$ is given to the leading order as $\varepsilon \to 0$ by parabola and constant 1, respectively. The maximum of $h_{0,\varepsilon}(x)$ is $O(1/\varepsilon)$, is attained at $x = 0$ and gives a so called "peak" of the droplet (see Figure 5.1). The value A/P is commonly called as the droplet half-width (see Glasner and Witelski [2]). The main result of the lemma given by assertion (ii) is that the ratio of the length of contact line region defined as $(a_\varepsilon/\varepsilon, b_\varepsilon/\varepsilon)$ to that one of whole interval $[-L/\varepsilon, L/\varepsilon]$ tends to 0 as $\varepsilon \to 0$. For the unscaled physical version of the no-slip equation(5.1) with (5.2) this means that the length of the contact line region $(a_\varepsilon, b_\varepsilon)$ tends to zero as $\varepsilon \to 0$. From the proof of the lemma in section 5.4 it is clear that the three intervals $(0, \ a_\varepsilon/\varepsilon)$, $(a_\varepsilon/\varepsilon, \ b_\varepsilon/\varepsilon)$, $(b_\varepsilon/\varepsilon, \ L/\varepsilon)$ with above asymptotical properties are not uniquely defined. In this chapter in order to escape from unnecessary technicalities we fix one possible definition for functions a_ε and b_ε. Once it is fixed then asymptotical bounds on the stationary $h_{0,\varepsilon}(x)$ stated in assertions (iii) and (iv) of the lemma are determined uniquely. This in turn determines asymptotical bounds for the functions $r_\varepsilon(x)$ and $f_\varepsilon(x)$ (see Lemma 5.17). ∎

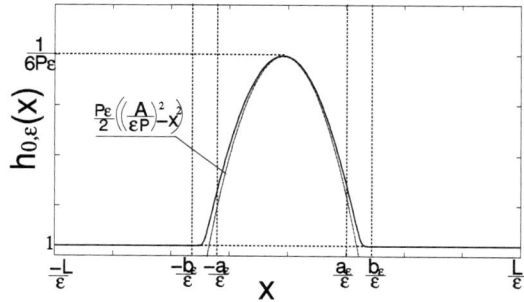

Figure 5.1: Stationary solution $h_{0,\varepsilon}(x)$ obtained numerically for $\varepsilon = 0.1$, $P = 0.1$, $L = 20$.

Next, let us state our main results (Theorems 5.10–5.12) about the asymptotics for the spectrum of EVP (5.18) as the singular parameter $\varepsilon \to 0$. Define a discrete countable set

$$M = \left\{ \left(\frac{\pi(2j+1)}{2(L-A/P)} \right)^2, \ j \in \mathbb{N}_0 \right\} \cup \{0\}, \tag{5.19}$$

with constant A given by (5.4). For a fixed $\varepsilon > 0$ denote also by σ_ε the spectrum of EVP (5.18).

Theorem 5.10. *(existence of a spectrum gap)*

(i) *If* $\{\lambda_l\} \in \sigma_{\varepsilon_l}$, $(l = 1, 2, ...)$ *is a sequence of eigenvalues to (5.18) corresponding to a sequence* $\{\varepsilon_l\} \to 0$ *and there exists a number* $K^* > 0$ *such that*

$$\left| \frac{\lambda_l}{\varepsilon_l^2} \right| \leq K^* \ \text{for all } l \in \mathbb{N}$$

5.2 Summary of Main Results and Discussion

then
$$\text{dist}\left(\frac{\lambda_l}{\varepsilon_l^2}, M\right) := \inf_{K \in M} \left| K - \frac{\lambda_l}{\varepsilon_l^2} \right| \to 0 \text{ as } l \to \infty.$$

(ii) For sufficiently small $\varepsilon > 0$ and any eigenvalue $\lambda \in \sigma_\varepsilon$ one has

$$\lambda \notin \left(0, \left[\frac{\pi}{4(L - A/P)}\varepsilon\right]^2\right). \tag{5.20}$$

Theorem 5.11. *(existence of eigenvalues with prescribed asymptotics)*
For every $j \in \mathbb{N}_0$ there exist positive numbers ε^j, δ^j and functions $\lambda_N^j, \lambda_D^j \in C^1((0, \varepsilon^j), \mathbb{R})$ such that for all $\varepsilon \in (0, \varepsilon^j)$ the following holds:

(i) $\lambda_N^j(\varepsilon) \in \sigma_\varepsilon$ and $\lambda_D^j(\varepsilon) \in \sigma_\varepsilon$,

(ii) $\left|\lambda_N^j(\varepsilon) - \left(\frac{\pi(2j+1)}{2(L - A/P)}\varepsilon\right)^2\right| = o_1(\varepsilon^2)$ and $\left|\lambda_D^j(\varepsilon) - \left(\frac{\pi(2j+1)}{2(L - A/P)}\varepsilon\right)^2\right| = o_2(\varepsilon^2)$,

(iii) If $\lambda \in \sigma_\varepsilon$ and $\left|\lambda - \left(\frac{\pi(2j+1)}{2(L - A/P)}\varepsilon\right)^2\right| \leq \delta^j \varepsilon^2$ then $\lambda = \lambda_N^j(\varepsilon)$ or $\lambda = \lambda_D^j(\varepsilon)$.

Let us for each $\varepsilon > 0$ denote by h_ε^- the minimum of stationary solution $h_{0,\varepsilon}$, which by (5.11b)–(5.11c) is attained in points $x = \pm L/\varepsilon$. From assertion (iii) of Theorem 5.4 it follows that any solution to EVP (5.18) solves classically (5.15a)–(5.15b). Vice versa any solution to (5.15a)–(5.15b) gives a solution to (5.18). Using (5.11a) and definitions (5.14a)–(5.14b) one can easily deduce that for each $\varepsilon > 0$ the pair $[h_{0,\varepsilon}(x) - h_\varepsilon^-, 0]$ satisfies (5.15a), but not (5.15b) because

$$h''_{0,\varepsilon}(\pm L/\varepsilon) \neq 0.$$

Indeed, if e.g. $h''_{0,\varepsilon}(L/\varepsilon) = 0$ then from $h_{0,\varepsilon}(L/\varepsilon) = h_\varepsilon^-$ and the fact that stationary solution $h_{0,\varepsilon}(x)$ satisfies boundary conditions (5.9) it follows that at the point $x = L/\varepsilon$ function $h_{0,\varepsilon}(x) - h_\varepsilon^-$ and its first three derivatives should be zero. Next, by uniqueness of solution to equation (5.15a) with $\lambda = 0$ and given initial condition $h^{(k)}(L/\varepsilon) = 0$ for $k = 0, 1, 2, 3$ it follows that $h_{0,\varepsilon}(x) - h_\varepsilon^- \equiv 0$. But this contradicts with the fact that for each $\varepsilon > 0$ stationary solution $h_{0,\varepsilon}(x)$ by its definition is not a constant. Consequently, using (5.11b) one concludes that $h''_{0,\varepsilon}(-L/\varepsilon) = h''_{0,\varepsilon}(L/\varepsilon) \neq 0$.

Nevertheless, in Lemma 5.23, section 5.4 we show that $h''_{0,\varepsilon}(\pm L/\varepsilon)$ tends to zero exponentially as $\varepsilon \to 0$. In view of above observations, naturally arises a question if there exists an eigenvalue of EVP (5.18) which exponentially tends to zero as $\varepsilon \to 0$. The next theorem answers this question.

Theorem 5.12. *(existence of exponentially small eigenvalue)*
There exist positive constants $c^*, \alpha, \varepsilon^*, \delta^*$ and function $\lambda^* \in C^1((0, \varepsilon^*), \mathbb{R})$ such that for all $\varepsilon \in (0, \varepsilon^*)$ the following holds:

(i) $\lambda^*(\varepsilon) \in \sigma_\varepsilon$

(ii) $|\lambda^*(\varepsilon)| \leq c^* \varepsilon^{1/2} \exp\left(-\frac{\alpha}{\varepsilon^{2/3}}\right)$,

(iii) If $\lambda \in \sigma_\varepsilon$ and $|\lambda| \leq \delta^* \varepsilon^2$ then $\lambda = \lambda^*(\varepsilon)$.

Chapter 5 Spectrum Asymptotics in a Singular Limit

We should point out that the form of exponential small term $\exp\left(-\frac{\alpha}{\varepsilon^{2/3}}\right)$ in the assertion (ii) of this theorem is strongly connected with estimate (5.47) which we obtain for $h''_{0,\varepsilon}(\pm L/\varepsilon)$ in Lemma 5.23.

Remark 5.13. The last three theorems together give a result which could be interesting for applications. Namely, in the spectrum of EVP (5.18) a set of positive eigenvalues $\mathcal{R}_\varepsilon := \{\lambda^j_D(\varepsilon), \lambda^j_N(\varepsilon)\ j \in \mathbb{N}_0\}$ is separated from exactly one exponentially small negative eigenvalue $\lambda^*(\varepsilon)$ by a spectrum gap given in (5.20). Note that the right end of it we choose as $K^1\varepsilon^2/4$ where K^1 is the smallest positive element of the set M from (5.19).

Remark 5.14. Elements of the above set $\mathcal{R}_\varepsilon$ have asymptotics of $O(\varepsilon^2)$. In this study we do not state any results on the existence of eigenvalues with asymptotics $\gg O(\varepsilon^2)$, but their possible presence by no-means influences the spectral gap property described above. ∎

Remark 5.15. Using relation (5.13) the results of Theorems 5.10–5.12 can be easily reformulated for the unscaled (physically relevant) EVP (5.5). We remain this to the reader and just want to point out that the spectrum gap analog of (5.20) in this case will be given by

$$-\lambda \notin \left(-\left[\frac{\pi}{4(L-A/P)}\right]^2 \varepsilon, 0\right).$$

∎

The structure for the rest of this chapter is as follows. In section 5.3 we decompose EVP (5.18) to two EVPs on the half-interval $[0, L/\varepsilon]$, which we call Dirichlet and Neumann half-droplet problems, respectively. In section 5.4 we prove Lemmata 5.8, 5.17 stating asymptotics for stationary solution $h_{0,\varepsilon}(x)$ and coefficient functions $r_\varepsilon(x)$ and $f_\varepsilon(x)$ of the symmetric EVP (5.18) as $\varepsilon \to 0$. In the sections 5.5-5.6 we describe two approximate problems the spectra of which are approximations from below and above for the spectrum of the Dirichlet half droplet problem in a sense stated in Proposition 5.20. The analogous analysis can be applied also to the Neumann half-droplet problem (see Remarks 5.16 and 5.25). In section 5.5 we prove Theorem 5.24, an analog of Theorem 5.10 for the approximate problems. In section 5.6 we prove Theorem 5.29, an analog of Theorem 5.11 for one of the approximate problems. To this end we apply a modified implicit function Theorem 5.30 first introduced in Magnus [41] and Recke and Omel'chenko [5]. In section 5.7 we prove Theorems 5.10–5.12 together. In section 5.8 we give a numerical confirmation of the spectral properties stated in Theorems 5.10–5.12.

5.3 Half-droplet Problem and its Approximations

By (5.11b) for each $\varepsilon > 0$ stationary solution $h_{0,\varepsilon}(x)$ is an even function. Hence functions (5.14a)–(5.14b) are also even. Therefore, if $[h(x), \lambda]$ is an eigenpair of EVP (5.18) then $[h(-x), \lambda]$ is also an eigenpair of it. If $h(x)$ is not an even or odd function, then the multiplicity of λ is at least two. Indeed, numerical solutions from section 5.8 give us pairs of very close eigenvalues, which indicate that there could be eigenvalues of (5.18) with multiplicity 2 (see also the formulation of Theorem 5.11). But for application of a modified implicit function theorem (Theorem 5.30) we would like to work with simple eigenvalues. To this end and also to simplify subsequent calculations we introduce a decomposition of (5.18) to two EVPs on the half-interval $[0, L/\varepsilon]$. Every eigenfunction $h(x)$ of (5.18) defines an eigensubspace which is spanned by even eigenfunction $h^e(x) := (h(x) + h(-x))/2$ and odd one $h^o(x) := (h(x) - h(-x))/2$ (one of them may be identically zero). This decomposition one can actually apply to any function in Hilbert space W_ε defined in (5.16). Therefore, one can decompose W_ε into a direct sum of the closed subspace of even functions W_ε^e and the closed subspace of odd functions W_ε^o:

$$W_\varepsilon = W_\varepsilon^e \oplus W_\varepsilon^o.$$

Analogously, any eigensubspace for EVP (5.18) can be decomposed in two, one of which belongs to W_ε^e and another to W_ε^o. Using this and again even property of functions $r_\varepsilon(x)$, $f_\varepsilon(x)$ one obtains that the set of solutions to EVP (5.18) is the union of the sets of solutions of two symmetric EVPs:

$$h \in W_\varepsilon^o, \lambda \in \mathbb{R} : \int_{-L/\varepsilon}^{L/\varepsilon} (h''w'' - r_\varepsilon h'w' - \lambda f_\varepsilon hw)\, dx = 0 \text{ for all } w \in W_\varepsilon^o,$$

and

$$h \in W_\varepsilon^e, \lambda \in \mathbb{R} : \int_{-L/\varepsilon}^{L/\varepsilon} (h''w'' - r_\varepsilon h'w' - \lambda f_\varepsilon hw)\, dx = 0 \text{ for all } w \in W_\varepsilon^e.$$

Moreover, it is easy to see that the first EVP above is equivalent to one called in this chapter as Dirichlet **half-droplet problem**:

$$h \in V_\varepsilon, \lambda \in \mathbb{R} : \int_0^{L/\varepsilon} (h''w'' - r_\varepsilon h'w' - \lambda f_\varepsilon hw)\, dx = 0 \text{ for all } w \in V_\varepsilon; \quad (5.21)$$

and the second EVP to one called in this chapter as Neumann **half-droplet problem**

$$h \in Q_\varepsilon, \lambda \in \mathbb{R} : \int_0^{L/\varepsilon} (h''w'' - r_\varepsilon h'w' - \lambda f_\varepsilon hw)\, dx = 0 \text{ for all } w \in Q_\varepsilon, \quad (5.22)$$

where Hilbert spaces V_ε and Q_ε are defined as

$$V_\varepsilon = H^2(0, L/\varepsilon) \cap H_0^1(0, L/\varepsilon),$$
$$Q_\varepsilon = \left\{ h \in H^2(0, L/\varepsilon) : h'(0) = h(L/\varepsilon) = 0 \right\}, \quad (5.23)$$

and both are equipped with the standard inner product of $H^2(0, L/\varepsilon)$.

***Remark* 5.16.** Below we introduce two approximate EVP problems for the Dirichlet half-droplet problem (5.21) and prove several results about their solutions in this and next two sections. In section 5.7 we will use the fact that analogous approximate problems can be defined for the Neumann half-droplet problem (5.22) and analogs of Propositions 5.18–5.20, Theorems 5.24, 5.29 and Lemma 5.27 can be proved for them in the exact the same manner. ∎

The next lemma, which proof is given in section 5.4, determines asymptotics for functions (5.14a)–(5.14b) as $\varepsilon \to 0$.

Lemma 5.17. *For sufficiently small $\varepsilon > 0$ the following holds:*

(i)
$$0 \leq r_\varepsilon(x) \leq 2\varepsilon^3, \text{ for } x \leq a_\varepsilon/\varepsilon, \quad -1 \leq r_\varepsilon(x) = -1 + O\left(\varepsilon^{1/6}\right), \text{ for } x \geq b_\varepsilon/\varepsilon;$$
$$\varepsilon^3 \leq f_\varepsilon(x) \leq \varepsilon^{9/4}, \quad \text{for } x \leq a_\varepsilon/\varepsilon, \quad 1 - O\left(\varepsilon^{1/6}\right) \leq f_\varepsilon(x) \leq 1, \quad \text{for } x \geq b_\varepsilon/\varepsilon.$$

(ii) There exists a unique point x_ε^m and a number $k_1 > 0$ such that $a_\varepsilon/\varepsilon < x_\varepsilon^m < b_\varepsilon/\varepsilon$ and $r_\varepsilon(x_\varepsilon^m) = k_1$ gives the maximum of $r_\varepsilon(x)$ on $[0, L/\varepsilon]$.

Chapter 5 Spectrum Asymptotics in a Singular Limit

(iii) The function $f_\varepsilon(x)$ monotonically increases on $[0, L/\varepsilon]$, and function $r_\varepsilon(x)$ is monotonically increases on $[0, x_\varepsilon^m]$ and decreases on $[x_\varepsilon^m, L/\varepsilon]$.

Next, we define four functions (see also Figure 5.2):

$$r_\varepsilon^1(x) = \begin{cases} 2\varepsilon^3, & 0 \leq x \leq a_\varepsilon/\varepsilon \\ k_1, & a_\varepsilon/\varepsilon < x \leq b_\varepsilon/\varepsilon \\ -1 + \varepsilon^{1/12}, & b_\varepsilon/\varepsilon < x \leq L/\varepsilon \end{cases}, \quad r_\varepsilon^2(x) = \begin{cases} 0, & 0 \leq x \leq a_\varepsilon/\varepsilon \\ -1, & a_\varepsilon/\varepsilon < x \leq L/\varepsilon \end{cases}; \quad (5.24a)$$

$$f_\varepsilon^1(x) = \begin{cases} \varepsilon^{9/4}, & 0 \leq x \leq a_\varepsilon/\varepsilon \\ 1, & a_\varepsilon/\varepsilon < x \leq L/\varepsilon \end{cases}, \quad f_\varepsilon^2(x) = \begin{cases} \varepsilon^4, & 0 \leq x \leq b_\varepsilon/\varepsilon \\ 1 - \varepsilon^{1/12}, & b_\varepsilon/\varepsilon < x \leq L/\varepsilon \end{cases}. \quad (5.24b)$$

where a_ε, b_ε and k_1 are defined in Lemmata 5.8, 5.17. Using (5.24a)–(5.24b) one can define for the Dirichlet half-droplet problem two approximate EVPs replacing functions (5.14a)-(5.14b) in (5.21) by their approximations $r_\varepsilon^i(x)$, $f_\varepsilon^i(x)$ with either $i = 1$ or $i = 2$.

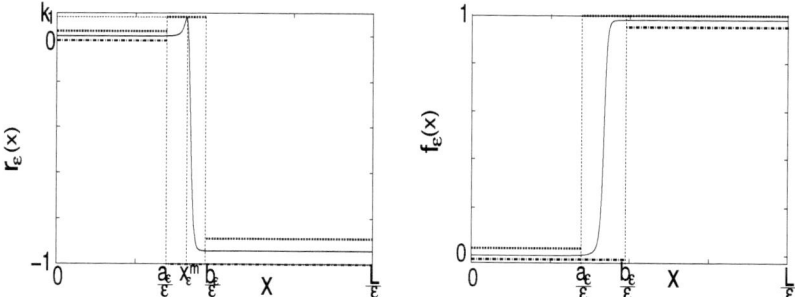

Figure 5.2: Function $r_\varepsilon(x)$ and its approximations (left), function $f_\varepsilon(x)$ and its approximations (right), obtained numerically for $L = 20$, $P = 0.1$ and $\varepsilon = 0.1$. Approximations $r_\varepsilon^1(x)$, $f_\varepsilon^1(x)$ are denoted by short dashes and $r_\varepsilon^2(x)$, $f_\varepsilon^2(x)$ are denoted by dash-dot lines.

Define now Hilbert space

$$H_\varepsilon = L^2(0, L/\varepsilon), \quad (5.25)$$

with an inner product

$$(h, \tilde{h})_{H_\varepsilon} := \int_0^{L/\varepsilon} h\tilde{h} f_\varepsilon^2 \, dx. \quad (5.26)$$

The next proposition is an analog of Theorem 5.4 for the approximate EVPs.

Proposition 5.18. *Consider two approximate EVPs,*

$$h \in V_\varepsilon, \, \lambda \in \mathbb{R} \, : \, \int_{-L/\varepsilon}^{L/\varepsilon} (h''w'' - r_\varepsilon^i h'w' - \lambda f_\varepsilon^i h\,w)\,dx = 0, \, \forall w \in V_\varepsilon. \quad (5.27)$$

with $i = 1$ or $i = 2$. For fixed i and $\varepsilon > 0$ there exist sequences $\{\lambda_\varepsilon^{i,j}\}$, $\{h_\varepsilon^{i,j}\}$, where $j \in \mathbb{N}_0$ such that:

(i) for each $j \in \mathbb{N}_0$ the pair $[h_\varepsilon^{i,j}, \lambda_\varepsilon^{i,j}]$ is a solution to (5.27);

5.3 Half-droplet Problem and its Approximations

(ii)
$$\lambda_\varepsilon^{i,0} \leq \lambda_\varepsilon^{i,1} \leq \lambda_\varepsilon^{i,2} \leq \ldots \to \infty; \tag{5.28}$$

(iii) the set $\{h_\varepsilon^{i,j}, \ j \in \mathbb{N}_0\}$ forms an orthonormal basis in (5.25) with respect to inner product (5.26).

The next proposition describes regularity for the solutions of (5.27) and introduces an important property of them, namely connection conditions (5.29)–(5.30b).

Proposition 5.19. Let $\varepsilon > 0$ be fixed and $[h, \lambda]$ be a solution to (5.27) for $i = 1$, $[\tilde{h}, \tilde{\lambda}]$ be a solution to (5.27) for $i = 2$. Then the following properties hold:

(i) On each of three intervals $(0, a_\varepsilon/\varepsilon)$, $(a_\varepsilon/\varepsilon, b_\varepsilon/\varepsilon)$ and $(b_\varepsilon/\varepsilon, L/\varepsilon)$
$$\tilde{h}^{(4)}(x) + \left(r_\varepsilon^2(x)\tilde{h}'(x)\right)' = \tilde{\lambda} f_\varepsilon^2(x)\tilde{h}(x) \text{ and } h^{(4)}(x) + \left(r_\varepsilon^1(x)h'(x)\right)' = \lambda f_\varepsilon^1(x)h(x).$$

(ii) At the point $x = b_\varepsilon/\varepsilon$ the function $\tilde{h}(x)$ is smooth, $h(x)$ is twice continuously differentiable and satisfies
$$h'''(b_\varepsilon/\varepsilon - 0) + k\, h'(b_\varepsilon/\varepsilon) = h'''(b_\varepsilon/\varepsilon + 0), \tag{5.29}$$
where $k := k_1 + 1 - \varepsilon^{1/12}$ and k_1 is defined in assertion (ii) of Lemma 5.17.

(iii) At the point $x = a_\varepsilon/\varepsilon$ both h and $\tilde{h}$ are twice continuously differentiable and satisfy:
$$\tilde{h}'''(a_\varepsilon/\varepsilon - 0) + \tilde{h}'(a_\varepsilon/\varepsilon) = \tilde{h}'''(a_\varepsilon/\varepsilon + 0). \tag{5.30a}$$
$$h'''(a_\varepsilon/\varepsilon - 0) - k_1\, h'(a_\varepsilon/\varepsilon) = h'''(a_\varepsilon/\varepsilon + 0). \tag{5.30b}$$

(iv) Both functions h and $\tilde{h}$ satisfy Dirichlet boundary conditions, namely
$$h''(0) = h(0) = h''(L/\varepsilon) = h(L/\varepsilon) = \tilde{h}''(0) = \tilde{h}(0) = \tilde{h}''(L/\varepsilon) = \tilde{h}(L/\varepsilon) = 0.$$

Proof: We prove assertions (i)–(iv) only for solutions $[\tilde{h}, \tilde{\lambda}]$. In the exact the same way the remaining assertions for $[h, \lambda]$ can be proved. From (5.27) with $i = 2$ it follows
$$\int_0^{a_\varepsilon/\varepsilon} (\tilde{h}''w'' - r_\varepsilon^2 \tilde{h}'w' - \tilde{\lambda} f_\varepsilon^2 \tilde{h}w)\, dx +$$
$$+ \int_{a_\varepsilon/\varepsilon}^{L/\varepsilon} (\tilde{h}''w'' - r_\varepsilon^2 \tilde{h}'w' - \tilde{\lambda} f_\varepsilon^2 \tilde{h}w)\, dx = 0, \ \forall w \in C_c^\infty(0, L/\varepsilon).$$

Integrating each integral in the last expression separately two times by parts and using definitions (5.24a)–(5.24b) gives
$$\int_0^{a_\varepsilon/\varepsilon} \left(\tilde{h}^{(4)} + \left(r_\varepsilon^2 \tilde{h}'\right)' - \tilde{\lambda} f_\varepsilon^2 \tilde{h}\right) w\, dx + \int_{a_\varepsilon/\varepsilon}^{L/\varepsilon} \left(\tilde{h}^{(4)} + \left(r_\varepsilon^2 \tilde{h}'\right)' - \tilde{\lambda} f_\varepsilon^2 \tilde{h}\right) w\, dx$$
$$+ \left(\tilde{h}''w'\right)\Big|_{a_\varepsilon/\varepsilon - 0}^{a_\varepsilon/\varepsilon + 0} - \left(\tilde{h}''' + r_\varepsilon^2 \tilde{h}'\right) w \Big|_{a_\varepsilon/\varepsilon - 0}^{a_\varepsilon/\varepsilon + 0} = 0, \ \forall w \in C_c^\infty(0, L/\varepsilon),$$

From this it follows that
$$\int_0^{a_\varepsilon/\varepsilon} \left(\tilde{h}^{(4)} + \left(r_\varepsilon^2 \tilde{h}'\right)' - \tilde{\lambda} f_\varepsilon^2 \tilde{h}\right) w\, dx = 0, \ \forall w \in C_c^\infty(0, a_\varepsilon/\varepsilon)$$
$$\int_{a_\varepsilon/\varepsilon}^{L/\varepsilon} \left(\tilde{h}^{(4)} + \left(r_\varepsilon^2 \tilde{h}'\right)' - \tilde{\lambda} f_\varepsilon^2 \tilde{h}\right) w\, dx = 0, \ \forall w \in C_c^\infty(a_\varepsilon/\varepsilon, L/\varepsilon).$$

Chapter 5 Spectrum Asymptotics in a Singular Limit

Hence assertion (i) for $[\tilde{h}, \tilde{\lambda}]$ is true and

$$\left(\tilde{h}''w'\right)\Big|_{a_\varepsilon/\varepsilon-0}^{a_\varepsilon/\varepsilon+0} - \left(\tilde{h}''' + r_\varepsilon^2 \tilde{h}'\right)w\Big|_{a_\varepsilon/\varepsilon-0}^{a_\varepsilon/\varepsilon+0} = 0, \; \forall w \in C_c^\infty(0, L/\varepsilon),$$

Taking in the last expression consequently test functions $w(x)$ such that $w'(a/\varepsilon) = 0$ or $w(a/\varepsilon) = 0$ the connection condition (5.30a) follows. The proof of assertion (iv) is completely analogous to that for the natural boundary conditions in Theorem 23 of Laugesen and Pugh [49]. ∎

The next proposition is a reformulation of minimax and monotonicity principles (see section I.6.10 of Kato [50]) for our case.

Proposition 5.20. *Let the spectra of approximate problems (5.27) with $i = 1, 2$ be given in the form (5.28) and analogously the spectrum of the Dirichlet half-droplet problem (5.21) be given in the form:*

$$\lambda_{D,\varepsilon}^0 \leq \lambda_{D,\varepsilon}^1 \leq \lambda_{D,\varepsilon}^2 \leq \ldots \to \infty. \tag{5.31}$$

Then for every $j \in \mathbb{N}_0$ and $i = 1, 2$ it holds

$$\lambda_{D,\varepsilon}^j = \max_{M_j} \min_{h \in M_j} \frac{\int_0^{L/\varepsilon} (h'')^2 - (h')^2 \, r_\varepsilon \, dx}{\int_0^{L/\varepsilon} h^2 f_\varepsilon \, dx},$$

$$\lambda_\varepsilon^{i,j} = \max_{M_j} \min_{h \in M_j} \frac{\int_0^{L/\varepsilon} (h'')^2 - (h')^2 \, r_\varepsilon^i \, dx}{\int_0^{L/\varepsilon} h^2 f_\varepsilon^i \, dx}, \tag{5.32}$$

where M_j is any subspace of codimension j of V_ε defined in (5.23). Moreover, for all $j \in \mathbb{N}_0$ and sufficiently small $\varepsilon > 0$ it holds

$$\lambda_\varepsilon^{1,j} \leq \lambda_{D,\varepsilon}^j \leq \lambda_\varepsilon^{2,j}. \tag{5.33}$$

Proof: The proof for (5.32) is the same as one from the section I.6.10 of Kato [50] for the minimax principle. Let us now prove (5.33). By Lemma 5.17 for sufficiently small ε and all $x \in [0, L/\varepsilon]$ it holds:

$$r_\varepsilon^2(x) \leq r_\varepsilon(x) \leq r_\varepsilon^1(x), \quad f_\varepsilon^2(x) \leq f_\varepsilon(x) \leq f_\varepsilon^1(x),$$

From this and (5.32) for all $j \in \mathbb{N}_0$ and sufficiently small $\varepsilon > 0$ relation (5.33) follows. ∎

Remark 5.21. Proposition 5.18 states that eigenvalues of the approximate EVPs (5.27) for $i = 1$ and $i = 2$ give approximations from below and above, respectively, for the corresponding eigenvalues of the Dirichlet eigenvalue problem (5.21). In the next sections one often explores relation (5.33). Below we call EVPs (5.27) for $i = 1$ and $i = 2$ as the approximate problems "from below" and "from above", respectively.

5.4 Asymptotics for Stationary Solutions

In this section we prove important Lemmata 5.8, 5.17. Let us consider equation

$$h''(x) = \Pi(h(x)) - \varepsilon P. \tag{5.34}$$

By results of Appendix A of Glasner and Witelski [29] (see also Figure 5.3) there exists a hyperbolic saddle point $\hat{h}_{sc,\varepsilon}^-$ and an elliptic center point $\hat{h}_{sc,\varepsilon}^c$ of equation (5.34), which are two real

5.4 Asymptotics for Stationary Solutions

roots of the algebraic equation $\Pi(h) - \varepsilon P = 0$ and have the following asymptotics:

$$\hat{h}^-_{sc,\varepsilon} = 1 + \varepsilon P + O(\varepsilon^2), \quad \hat{h}^c_{sc,\varepsilon} \sim (\varepsilon P)^{-1/3}. \tag{5.35}$$

Corresponding to this there exists a homoclinic solution $\hat{h}_{sc,\varepsilon}(x)$ to (5.34), minimum of which is given by $\hat{h}^-_{sc,\varepsilon}$, and its maximum $\hat{h}^+_{sc,\varepsilon}$ (see Glasner and Witelski [29]) has asymptotics

$$\hat{h}^+_{sc,\varepsilon} = \frac{1}{6\varepsilon P} + 1 + O(\varepsilon). \tag{5.36}$$

One can define a first integral for $\hat{h}_{sc,\varepsilon}(x)$ as

$$1/2\left(\hat{h}'_{sc,\varepsilon}(x)\right)^2 + \mathcal{U}_\varepsilon\left(\hat{h}_{sc,\varepsilon}(x)\right) = 0,$$

where

$$\mathcal{U}_\varepsilon(h) := -U(h) + U\left(\hat{h}^-_{sc,\varepsilon}\right) + \varepsilon P(h - \hat{h}^-_{sc,\varepsilon}), \tag{5.37a}$$

$$\mathcal{U}_\varepsilon\left(\hat{h}^-_{sc,\varepsilon}\right) = \mathcal{U}'_\varepsilon\left(\hat{h}^-_{sc,\varepsilon}\right) = \mathcal{U}_\varepsilon\left(\hat{h}^+_{sc,\varepsilon}\right) = 0. \tag{5.37b}$$

The function $U(h)$ in (5.37a) (see its plot in Figure 5.4 is such that $dU/dh = \Pi(h)$ and

$$U(h) := \frac{1}{3h^3} - \frac{1}{2h^2}. \tag{5.38}$$

The next proposition is used in the proof of Lemma 5.8 below.

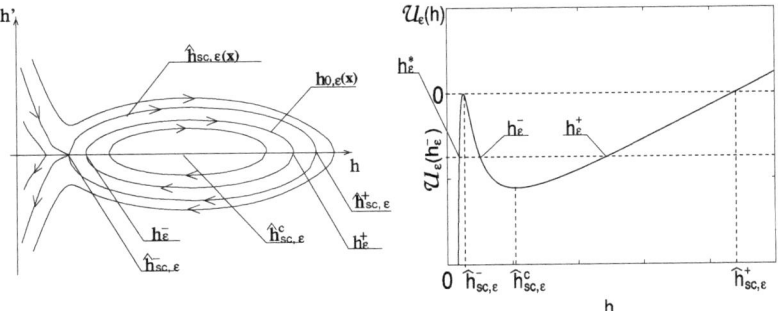

Figure 5.3: Phase plane portrait for the equation (5.34) (left) and plot of function $\mathcal{U}_\varepsilon(h)$ (right).

Proposition 5.22. *For each sufficiently small $\varepsilon > 0$ and $\delta \in (0, -\mathcal{U}_\varepsilon(\hat{h}^c_{sc,\varepsilon}))$ there exists a unique number $h_\varepsilon(\delta) \in (\hat{h}^-_{sc,\varepsilon}, \hat{h}^c_{sc,\varepsilon})$ such that*

$$\delta = -\mathcal{U}_\varepsilon(h_\varepsilon(\delta)).$$

Moreover, there exist positive numbers $\tilde{\varepsilon}, \tilde{\delta}$ such that for all $\varepsilon \in (0, \tilde{\varepsilon})$ and $\delta \in (0, \tilde{\delta})$ one has

$$h_\varepsilon(\delta) - \hat{h}^-_{sc,\varepsilon} < 2\sqrt{\delta}. \tag{5.39}$$

Proof: The existence and uniqueness of $h_\varepsilon(\delta)$ follows from the fact that $\hat{h}^-_{sc,\varepsilon}$ and $\hat{h}^c_{sc,\varepsilon}$ are double zero and local minimum of $\mathcal{U}_\varepsilon(h)$ (see also Figure 5.3). Moreover, using (5.37a)–(5.37b) and Peano formula one obtains

$$h_\varepsilon(\delta) - \hat{h}^-_{sc,\varepsilon} = \sqrt{\frac{2\delta}{\Pi'(\theta_\varepsilon(\delta))}}, \tag{5.40}$$

where $\theta_\varepsilon(\delta) \in [\hat{h}^-_{sc,\varepsilon}, h_\varepsilon(\delta)]$. From (5.8) it follows that that function $\Pi'(h)$ monotonically decreases for $h \in [1, 5/3]$. By (5.35) one has $\Pi'(\hat{h}^-_{sc,\varepsilon}) \to 1$ as $\varepsilon \to 0$ and $\Pi'(\hat{h}^c_{sc,\varepsilon}) < 0$ for sufficiently small $\varepsilon > 0$. Therefore, one can choose sufficiently small $\tilde{\varepsilon} > 0$ and $\tilde{\delta} > 0$ such that

$$\Pi'(h_{\tilde{\varepsilon}}(\tilde{\delta})) := 1/2. \tag{5.41}$$

Next, from (5.37a) and definition of $\hat{h}^-_{sc,\varepsilon}$ it follows that

$$\frac{\partial \mathcal{U}_\varepsilon(h)}{\partial \varepsilon} = (\Pi(\hat{h}^-_{sc,\varepsilon}) - \varepsilon\, P) \frac{\partial \hat{h}^-_{sc,\varepsilon}}{\partial \varepsilon} + P(h - \hat{h}^-_{sc,\varepsilon}) =$$
$$= P(h - \hat{h}^-_{sc,\varepsilon}) > 0 \text{ for all } h > \hat{h}^-_{\tilde{\varepsilon}} \text{ and } \varepsilon \in (0, \tilde{\varepsilon}). \tag{5.42}$$

Let us fix any $0 < \varepsilon < \tilde{\varepsilon}$ and $0 < \delta < \tilde{\delta}$. Define a number $h^* > \hat{h}^-_{sc,\tilde{\varepsilon}}$ such that $\mathcal{U}_{\tilde{\varepsilon}}(h^*) := -\delta$. If we suppose that $h^* \leq h_\varepsilon(\delta)$ then we arrive to the following contradiction:

$$-\delta = \mathcal{U}_{\tilde{\varepsilon}}(h^*) \geq \mathcal{U}_{\tilde{\varepsilon}}(h_\varepsilon(\delta)) > \mathcal{U}_\varepsilon(h_\varepsilon(\delta)) = -\delta,$$

where we used (5.42) and that function $\mathcal{U}_{\tilde{\varepsilon}}(h)$ decreases for $h \in (\hat{h}^-_{sc,\tilde{\varepsilon}}, \hat{h}^c_{sc,\tilde{\varepsilon}})$. Therefore, $h^* > h_\varepsilon(\delta)$. On the other hand

$$\mathcal{U}_{\tilde{\varepsilon}}(h^*) = -\delta > -\tilde{\delta} = \mathcal{U}_{\tilde{\varepsilon}}(h_{\tilde{\varepsilon}}(\tilde{\delta}))$$

and therefore again by monotonicity of $\mathcal{U}_{\tilde{\varepsilon}}(h)$ one gets $h^* < h_{\tilde{\varepsilon}}(\tilde{\delta})$. Combining all together one obtains $h_\varepsilon(\delta) < h_{\tilde{\varepsilon}}(\tilde{\delta})$. Finally, using again monotonicity of $\Pi'(h)$ and definition (5.41) one obtains that $\Pi'(h_\varepsilon(\delta)) > 1/2$, and therefore from (5.40) estimate (5.39) follows. ∎

Next, we define h^+_ε and h^-_ε as the maximum and the minimum of the stationary solution $h_{0,\varepsilon}(x)$ which are attained at $x = 0$ and $x = \pm L/\varepsilon$ by (5.11b)–(5.11c). A first integral for $h_{0,\varepsilon}(x)$ is determined by

$$1/2(h'_{0,\varepsilon}(x))^2 + \mathcal{U}_\varepsilon(h_{0,\varepsilon}(x)) - \mathcal{U}_\varepsilon(h^-_\varepsilon) = 0, \tag{5.43a}$$
$$\mathcal{U}_\varepsilon(h^+_\varepsilon) - \mathcal{U}_\varepsilon(h^-_\varepsilon) = 0 \tag{5.43b}$$

In the next lemma we state asymptotics for h^+_ε, h^-_ε and $h''_{0,\varepsilon}(\pm L/\varepsilon)$ as $\varepsilon \to 0$. We should point out that an appearance of the term $\frac{\alpha}{\varepsilon^{2/3}}$ in estimates (5.44) and (5.47) is strongly connected with the fact that by asymptotics (5.35), (5.36)

$$\frac{\hat{h}^+_{sc,\varepsilon}}{\hat{h}^c_{sc,\varepsilon}} = O(\varepsilon^{-2/3}).$$

Lemma 5.23. *There exist a positive constant α such that for all sufficiently small $\varepsilon > 0$ it holds*

(i)

$$h^-_\varepsilon - \hat{h}^-_{sc,\varepsilon} \leq \exp\left(-\frac{\alpha}{\varepsilon^{2/3}}\right), \tag{5.44}$$

5.4 Asymptotics for Stationary Solutions

(ii)
$$h_\varepsilon^- = 1 + \varepsilon P + o(\varepsilon), \tag{5.45}$$

(iii)
$$h_\varepsilon^+ \sim \frac{1}{6\varepsilon P}. \tag{5.46}$$

(iv)
$$\left| h_{0,\varepsilon}''(\pm L/\varepsilon) \right| \leq \exp\left(-\frac{\alpha}{\varepsilon^{2/3}}\right), \tag{5.47}$$

Proof: a) Integrating (5.43a) with respect to x on $(0, L/\varepsilon)$ and using $h_{0,\varepsilon}(0) = h_\varepsilon^+$, $h_{0,\varepsilon}(L/\varepsilon) = h_\varepsilon^-$ one obtains
$$\frac{L}{\varepsilon} = \int_{h_\varepsilon^-}^{h_\varepsilon^+} \frac{dh}{\sqrt{2\left(\mathcal{U}_\varepsilon(h_\varepsilon^-) - \mathcal{U}_\varepsilon(h)\right)}}. \tag{5.48}$$

From (5.37a) and (5.43b) one obtains
$$\mathcal{U}_\varepsilon(h_\varepsilon^-) - \mathcal{U}_\varepsilon(h) = U(h) - U(h_\varepsilon^-) - \varepsilon P(h - h_\varepsilon^-), \tag{5.49a}$$
$$U(h_\varepsilon^+) - U(h_\varepsilon^-) - \varepsilon P(h_\varepsilon^+ - h_\varepsilon^-) = 0. \tag{5.49b}$$

By (5.37a), (5.38) for a fixed $\varepsilon > 0$ function $\mathcal{U}_\varepsilon(h)$ monotonically increases on $(0, \hat{h}_{sc,\varepsilon}^-)$ from $-\infty$ to 0, decreases on $(\hat{h}_{sc,\varepsilon}^-, \hat{h}_{sc,\varepsilon}^c)$ and increases on $(\hat{h}_{sc,\varepsilon}^c, \hat{h}_{sc,\varepsilon}^+)$ (see Figure 5.3). Using this and (5.49a) one arrives to the following representation:
$$\mathcal{U}_\varepsilon(h_\varepsilon^-) - \mathcal{U}_\varepsilon(h) = \frac{\varepsilon P(h - h_\varepsilon^-)(h - h_\varepsilon^*)(h - h_\varepsilon^{**})(h_\varepsilon^+ - h)}{h^3}, \tag{5.50}$$

where four real zeros of function $\mathcal{U}_\varepsilon(h_\varepsilon^-) - \mathcal{U}_\varepsilon(h)$ for each fixed $\varepsilon > 0$ fulfill the following constraints:
$$h_\varepsilon^{**} < 0, \ 0 < h_\varepsilon^* < \hat{h}_{sc,\varepsilon}^-,$$
$$\hat{h}_{sc,\varepsilon}^- < h_\varepsilon^- < \hat{h}_{sc,\varepsilon}^c, \ \hat{h}_{sc,\varepsilon}^c < h_\varepsilon^+ < \hat{h}_{sc,\varepsilon}^+. \tag{5.51}$$

b) Let us prove using a contradiction argument that there exists positive numbers ε_1 and α_1 such that
$$h_\varepsilon^- \leq \alpha_1 \text{ for all } \varepsilon \in (0, \varepsilon_1). \tag{5.52}$$

Suppose inverse then without loss of generality $h_\varepsilon^- \to +\infty$ as $\varepsilon \to 0$. Using (5.48), (5.50) and (5.51) one estimates
$$\frac{L}{\varepsilon} \leq \frac{1}{\sqrt{2\varepsilon P}} \int_{h_\varepsilon^-}^{h_\varepsilon^+} \frac{\sqrt{h_\varepsilon^+}}{\sqrt{(h - h_\varepsilon^-)(h_\varepsilon^+ - h)}} \sqrt{\frac{h}{h - \hat{h}_{sc,\varepsilon}^-}} \, dh.$$

From $h_\varepsilon^- \to +\infty$ and asymptotics (5.35) it follows that there exists $\tilde{\varepsilon} > 0$ such that
$$\sqrt{\frac{h}{h - \hat{h}_{sc,\varepsilon}^-}} \leq \sqrt{\frac{h_\varepsilon^-}{h_\varepsilon^- - \hat{h}_{sc,\tilde{\varepsilon}}^-}} \leq \sqrt{2}$$

Chapter 5 Spectrum Asymptotics in a Singular Limit

for all $h \in (h_\varepsilon^-, +\infty)$ and $\varepsilon \in (0, \tilde{\varepsilon})$. Using last two estimates one obtains for all $\varepsilon \in (0, \tilde{\varepsilon})$

$$\frac{L}{\varepsilon} \leq \frac{1}{\sqrt{\varepsilon P}} \int_{h_\varepsilon^-}^{h_\varepsilon^+} \frac{\sqrt{h_\varepsilon^+}}{\sqrt{(h - h_\varepsilon^-)(h_\varepsilon^+ - h)}} dh = \sqrt{\frac{h_\varepsilon^+}{\varepsilon P}} \pi.$$

On the other hand by $h_\varepsilon^- \to +\infty$ and (5.49b) one has $\varepsilon P h_\varepsilon^+ = o(1)$ as $\varepsilon \to 0$. Using this and the last estimate one obtains that

$$\frac{L}{\varepsilon} \leq \frac{o(1)}{\varepsilon P} \pi,$$

which obviously gives a contradiction. Therefore, (5.52) holds with some positive numbers ε_1, α_1. Next, let us show using a contradiction argument that there exists positive numbers α_2 and ε_2 such that

$$h_\varepsilon^+ \geq \frac{\alpha_2}{\varepsilon} \text{ for all } \varepsilon \in (0, \varepsilon_2). \tag{5.53}$$

Suppose inverse then without loss of generality $\varepsilon P h_\varepsilon^+ \to 0$ as $\varepsilon \to 0$. On the other hand (5.51)

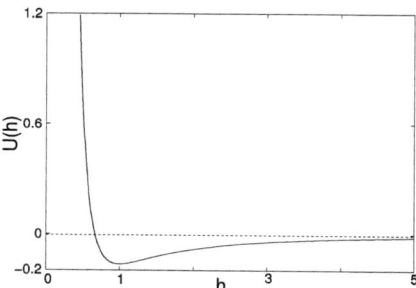

Figure 5.4: Plot of function $U(h)$.

and asymptotics (5.35) yield $h_\varepsilon^+ \to +\infty$ as $\varepsilon \to 0$. Substituting all this in (5.49b) and using (5.52), (5.38) one obtains

$$U(h_\varepsilon^-) \to 0.$$

From this by $h_\varepsilon^- > \hat{h}_{sc,\varepsilon}^- > 1$ and (5.38) it follows that $h_\varepsilon^- \to +\infty$, which gives a contradiction to (5.52). Therefore, estimate (5.53) holds with some positive numbers ε_2, α_2.

c) We write now formula (5.48) as

$$\frac{L}{\varepsilon} = I_1 + I_2 := \int_{h_\varepsilon^-}^{\hat{h}_{sc,\varepsilon}^c} \frac{dh}{\sqrt{2\left(\mathcal{U}_\varepsilon(h_\varepsilon^-) - \mathcal{U}_\varepsilon(h)\right)}} + \int_{\hat{h}_{sc,\varepsilon}^c}^{h_\varepsilon^+} \frac{dh}{\sqrt{2\left(\mathcal{U}_\varepsilon(h_\varepsilon^-) - \mathcal{U}_\varepsilon(h)\right)}} \tag{5.54}$$

and estimate each of the integrals I_k, $k = 1, 2$ separately. Using again (5.51) and (5.52) one estimates

$$I_2 = \int_{\hat{h}_{sc,\varepsilon}^c}^{h_\varepsilon^+} \frac{dh}{\sqrt{2\left(\mathcal{U}_\varepsilon(h_\varepsilon^-) - \mathcal{U}_\varepsilon(h)\right)}} \leq \frac{1}{\sqrt{2\varepsilon P}} \int_{\hat{h}_{sc,\varepsilon}^c}^{h_\varepsilon^+} \frac{h}{\sqrt{(h - \alpha_1)(h - \hat{h}_{sc,\varepsilon}^-)(h_\varepsilon^+ - h)}} dh.$$

By (5.35) and (5.51) both $\hat{h}_{sc,\varepsilon}^c$ and h_ε^+ tend to $+\infty$ and $\hat{h}_{sc,\varepsilon}^- \to 1$ as $\varepsilon \to 0$. Therefore, there

5.4 Asymptotics for Stationary Solutions

exists $\varepsilon^* > 0$ such that

$$\frac{h}{\sqrt{(h-\alpha_1)(h-\hat{h}_{sc,\varepsilon}^-)}} \leq (1+C) \text{ with } C := \frac{LP}{2A} - 1/2 \qquad (5.55)$$

holds for all $h \in (\hat{h}_{sc,\varepsilon^*}^c, +\infty)$ and $\varepsilon \in (0, \varepsilon^*)$. Note that number C in (5.55) by condition (5.4) is positive. Using now last two estimates, again (5.51) and asymptotics (5.36) one obtains

$$I_2 \leq \frac{1+C}{\sqrt{2\varepsilon P}} \int_{\hat{h}_{sc,\varepsilon}^c}^{h_\varepsilon^+} \frac{dh}{\sqrt{h_\varepsilon^+ - h}} \leq$$

$$\leq \sqrt{\frac{2\hat{h}_{sc,\varepsilon}^+}{\varepsilon P}}(1+C) = \left(\frac{L+A/P}{2}\right)\frac{1}{\varepsilon} + o\left(\frac{1}{\varepsilon}\right). \qquad (5.56)$$

d) Let us now estimate integral I_1 from (5.54). Using again (5.51) one obtains

$$I_1 = \int_{h_\varepsilon^-}^{\hat{h}_{sc,\varepsilon}^c} \frac{dh}{\sqrt{2\left(\mathcal{U}_\varepsilon(h_\varepsilon^-) - \mathcal{U}_\varepsilon(h)\right)}} \leq$$

$$\leq \frac{1}{\sqrt{2\varepsilon P}} \frac{\hat{h}_{sc,\varepsilon}^c}{\sqrt{h_\varepsilon^+ - \hat{h}_{sc,\varepsilon}^c}} \int_{h_\varepsilon^-}^{\hat{h}_{sc,\varepsilon}^c} \frac{dh}{\sqrt{(h-h_\varepsilon^-)(h-\hat{h}_{sc,\varepsilon}^-)}}.$$

This, asymptotics (5.53) and (5.35) yield that there exist positive numbers ε_3 and α_3 such that for all $\varepsilon \in (0, \varepsilon_3)$ the following estimate holds:

$$I_1 \leq \frac{\alpha_3}{\varepsilon^{1/3}} \int_{h_\varepsilon^-}^{\hat{h}_{sc,\varepsilon}^c} \frac{dh}{\sqrt{(h-h_\varepsilon^-)(h-\hat{h}_{sc,\varepsilon}^-)}} \leq$$

$$\leq \frac{\alpha_3}{\varepsilon^{1/3}} \left(-\log\left(h_\varepsilon^- - \hat{h}_{sc,\varepsilon}^-\right) + 2\log\left(2\left(\hat{h}_{sc,\varepsilon}^c - \hat{h}_{sc,\varepsilon}^-\right)\right)\right). \qquad (5.57)$$

Finally, combining estimates (5.56), (5.57) together with formula (5.54) and using asymptotics (5.35) one obtains

$$\log(h_\varepsilon^- - \hat{h}_{sc,\varepsilon}^-) \leq -\left(\frac{L-A/P}{2\alpha_3}\right)\frac{1}{\varepsilon^{2/3}} + o\left(\frac{1}{\varepsilon^{2/3}}\right).$$

This together with condition (5.4) imply that estimate (5.44) holds for sufficiently small $\varepsilon > 0$ with positive constant

$$\alpha := \frac{L-A/P}{4\alpha_3}.$$

e) Asymptotics (5.45) follows from (5.44) and (5.35). In turn asymptotics (5.46) follows from (5.45) and (5.49b), (5.38).

f) Let us finally show estimate (5.47). Using (5.11a) and Peano formula one obtains

$$h''_{0,\varepsilon}(\pm L) = \Pi\left(h_\varepsilon^-\right) - \varepsilon P = \Pi'(\hat{h}_{sc,\varepsilon}^-)(h_\varepsilon^- - \hat{h}_{sc,\varepsilon}^-) + \Pi''(\theta_\varepsilon)(h_\varepsilon^- - \hat{h}_{sc,\varepsilon}^-)^2,$$

where $\theta_\varepsilon \in [\hat{h}_{sc,\varepsilon}^-, h_\varepsilon^-]$. Asymptotics (5.35) and (5.45) yield that $\Pi''(\theta_\varepsilon) < 0$ and $\Pi'(\hat{h}_{sc,\varepsilon}^-) < 1$ for sufficiently small $\varepsilon > 0$. Therefore, applying estimate (5.44) one obtains (5.47). ∎

Let us now prove Lemmata 5.8, 5.17.

Proof of Lemma 5.8: Using estimates (5.45)–(5.46) and (5.11c) for each sufficiently small

Chapter 5 Spectrum Asymptotics in a Singular Limit

$\varepsilon > 0$ define uniquely a_ε as

$$L > a_\varepsilon > 0: \quad h_{0,\varepsilon}(a_\varepsilon/\varepsilon) := \varepsilon^{-3/4} \tag{5.58}$$

Then by (5.8) and (5.58) one has

$$\Pi(h_{0,\varepsilon}(x)) = O(\varepsilon^{9/4}) \quad \text{and} \quad \frac{d^2 h_{0,\varepsilon}(x)}{dx^2} = -P\varepsilon + O(\varepsilon^{9/4}), \text{ for } x \in [0, a_\varepsilon/\varepsilon].$$

Integrating two times and using $h'_{0,\varepsilon}(0) = 0$ by (5.11c) one obtains

$$h'_{0,\varepsilon}(x) \sim -P\varepsilon x,$$
$$h_{0,\varepsilon}(x) \sim \frac{P\varepsilon}{2}\left(\left(\frac{C}{\varepsilon}\right)^2 - x^2\right) \quad \text{for all } x \in [0, a_\varepsilon/\varepsilon].$$

Taking $x = 0$ in the last expression and using (5.46) one obtains

$$C = \frac{A}{P}.$$

Taking next $x = a_\varepsilon$ gives

$$a_\varepsilon \sim \frac{A}{P} \quad \text{and} \quad h'_{0,\varepsilon}(a_\varepsilon/\varepsilon) \sim -A. \tag{5.59}$$

The estimate $\varepsilon^{-3/4} \leq h_{0,\varepsilon}(x) = O(1/\varepsilon)$ for all $x \in [0, a_\varepsilon/\varepsilon]$ follows from asymptotics (5.46), definition (5.58) and monotonicity of $h_{0,\varepsilon}(x)$ for $x > 0$ by (5.11c). Therefore, assertion (iii) of the lemma is proved.

Next, for each $\varepsilon > 0$ by definition of $\hat{h}^c_{sc,\varepsilon}$ there exists a unique $x^c_\varepsilon \in (0, L)$ such that

$$h_{0,\varepsilon}(x^c_\varepsilon/\varepsilon) = \hat{h}^c_{sc,\varepsilon} \quad \text{and} \quad h''_{0,\varepsilon}(x^c_\varepsilon/\varepsilon) = 0. \tag{5.60}$$

For each $\varepsilon > 0$ function $h_{0,\varepsilon}(x)$ decreases on $[0, L/\varepsilon]$, therefore one can define its inverse function $x_\varepsilon(h)$ decreasing on $[h^-_\varepsilon, h^+_\varepsilon]$ as

$$x_\varepsilon(h_{0,\varepsilon}(x)) := x \quad \text{and} \quad x'_\varepsilon(h) = \frac{1}{h'_{0,\varepsilon}(x_\varepsilon(h))}. \tag{5.61}$$

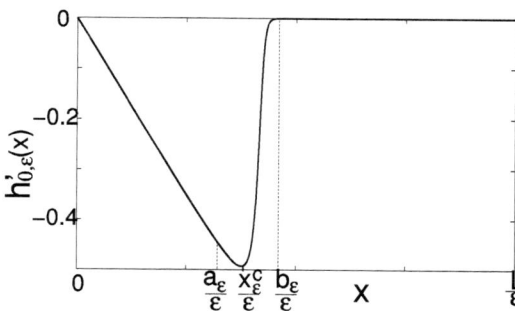

Figure 5.5: Plot of $h'_{0,\varepsilon}(x)$ obtained numerically for $\varepsilon = 0.1$, $P = 0.1$, $L = 20$ and corresponding to $h_{0,\varepsilon}(x)$ from Figure 5.1.

5.4 Asymptotics for Stationary Solutions

By this and (5.35), (5.58) and (5.59) one obtains

$$\frac{|x_\varepsilon^c - a_\varepsilon|}{\varepsilon} \leq \int_0^1 \left| x_\varepsilon' \left(t\,\hat{h}_{sc,\varepsilon}^c - (1-t)\varepsilon^{-3/4} \right) \right| dt \left| \hat{h}_{sc,\varepsilon}^c - \varepsilon^{-3/4} \right| \leq$$

$$\max_{\hat{h}_{sc,\varepsilon}^c \leq h \leq \varepsilon^{-3/4}} |x_\varepsilon'(h)| \left| \hat{h}_{sc,\varepsilon}^c - \varepsilon^{-3/4} \right| = \frac{\left| \hat{h}_{sc,\varepsilon}^c - \varepsilon^{-3/4} \right|}{|h_{0,\varepsilon}'(a_\varepsilon/\varepsilon)|} = O(\varepsilon^{-3/4}),$$

where we also use that $\left|h_{0,\varepsilon}'(x)\right|$ increases for $x \in [0,\, x_\varepsilon^c/\varepsilon]$ by (5.11a) and (5.60) (see also Figure 5.5). Therefore, using (5.59) one obtains

$$x_\varepsilon^c \sim a_\varepsilon \sim \frac{A}{P},$$
$$h_{0,\varepsilon}'(x_\varepsilon^c/\varepsilon) \sim -A,$$
$$x_\varepsilon^c - a_\varepsilon = O(\varepsilon^{1/4}). \tag{5.62}$$

Using this and that $\left|h_{0,\varepsilon}'(x)\right|$ decreases for $x \in [x_\varepsilon^c/\varepsilon,\, L/\varepsilon]$ (see Figure 5.5), we define for each sufficiently small $\varepsilon > 0$ a unique b_ε as

$$L > b_\varepsilon > x_\varepsilon^c : \quad h_{0,\varepsilon}'(b_\varepsilon/\varepsilon) := -\varepsilon^{1/6}. \tag{5.63}$$

Using this, first integral (5.43a), definition (5.37a)–(5.37b), Peano formula and asymptotics (5.44) one obtains

$$-\mathcal{U}_\varepsilon(h_{0,\varepsilon}(b_\varepsilon/\varepsilon)) = \frac{1}{2}\varepsilon^{1/3} - \mathcal{U}_\varepsilon(h_\varepsilon^-) + \mathcal{U}_\varepsilon(\hat{h}_{sc,\varepsilon}^-) =$$

$$= \frac{1}{2}\left(\varepsilon^{1/3} + \Pi'(\theta_\varepsilon)(h_\varepsilon^- - \hat{h}_{sc,\varepsilon}^-)^2 \right) \leq \frac{1}{2}\left(\varepsilon^{1/3} + \exp\left(-\frac{\alpha}{\varepsilon^{2/3}}\right) \right) =$$

$$= O(\varepsilon^{1/3}),$$

where $\theta_\varepsilon \in [\hat{h}_{sc,\varepsilon}^-,\, h_\varepsilon^-]$ and by (5.35), (5.45) $\Pi'(\theta_\varepsilon) \leq \Pi'(\hat{h}_{sc,\varepsilon}^-) \leq 1$. Applying Proposition 5.22 to the last inequality one gets

$$h_{0,\varepsilon}(b_\varepsilon/\varepsilon) = 1 + O(\varepsilon^{1/6}) \tag{5.64}$$

Now, the assertion (iv) of the lemma follows from the last asymptotics, (5.45) and (5.11c).

Finally, we show $b_\varepsilon - a_\varepsilon = O(\varepsilon^{1/4})$. Using the inverse function $x_\varepsilon(h)$ defined in (5.61) one writes

$$\frac{|x_\varepsilon^c - b_\varepsilon|}{\varepsilon} \leq \int_0^1 \left| x_\varepsilon' \left(t\,\hat{h}_{sc,\varepsilon}^c - (1-t)h_{0,\varepsilon}(b_\varepsilon/\varepsilon) \right) \right| dt \left| \hat{h}_{sc,\varepsilon}^c - h_{0,\varepsilon}(b_\varepsilon/\varepsilon) \right| \leq$$

$$\leq \max_{h_{0,\varepsilon}(b_\varepsilon/\varepsilon) \leq h \leq \hat{h}_{sc,\varepsilon}^c} |x_\varepsilon'(h)| \left| \hat{h}_{sc,\varepsilon}^c - O(1) \right| = \frac{\left| \hat{h}_{sc,\varepsilon}^c - O(1) \right|}{\varepsilon^{1/6}} = O(\varepsilon^{-1/2}),$$

where we also used definition (5.63) and asymptotics (5.64), (5.35). From the last estimate one obtains $b_\varepsilon - x_\varepsilon^c = O(\varepsilon^{1/2})$. Combining this with (5.62) yields $b_\varepsilon - a_\varepsilon = O(\varepsilon^{1/4})$, which in turn noting (5.59) implies assertion (ii) of the lemma. This concludes the proof of the lemma. ∎

Proof of Lemma 5.17: By definition (5.14a)

$$r_\varepsilon(x) = -\frac{4}{(h_{0,\varepsilon}(x))^5} + \frac{3}{(h_{0,\varepsilon}(x))^4} \tag{5.65}$$

By assumption (iii) of Lemma 5.8 and (5.65), (5.14b) it follows

$$O(\varepsilon^4) \leq r_\varepsilon(x) = \varepsilon^3 + o(\varepsilon^3), \quad O(\varepsilon^3) \leq f_\varepsilon(x) \leq \varepsilon^{9/4} \text{ for all } x \in [0,\, a_\varepsilon/\varepsilon]$$

By assumption (iv) of Lemma 5.8 and (5.65), (5.14b) it follows

$$-1 \leq r_\varepsilon(x) = -1 + O(\varepsilon^6), \quad 1 - O(\varepsilon^6) \leq f_\varepsilon(x) \leq 1 \text{ for all } x \in [b_\varepsilon/\varepsilon,\, L/\varepsilon].$$

Therefore, assertion (i) follows. Stationary points of the function $r_\varepsilon(x)$ are given by equation

$$r'_\varepsilon(x) = \left(\frac{20}{(h_{0,\varepsilon}(x))^6} - \frac{12}{(h_{0,\varepsilon}(x))^5}\right) h'_{0,\varepsilon}(x) = 0$$

Using asymptotics (5.45)–(5.46) and that $h'_{0,\varepsilon}(x) < 0$ for all $x \in (0,\, L/\varepsilon)$ one obtains that for each sufficiently small $\varepsilon > 0$ there exists a unique x^m_ε at which $r_\varepsilon(x)$ attains its maximum $k_1 := (3/5)^5 > 0$ such that $h_{0,\varepsilon}(x^m_\varepsilon) = 5/3$. Therefore, by (5.58), (5.64) and again $h'_{0,\varepsilon}(x) < 0$ for all $x \in (0,\, L/\varepsilon)$ it follows that $a_\varepsilon/\varepsilon < x^m_\varepsilon < b_\varepsilon/\varepsilon$, and hence the assertions (ii) and (iii) are proved. ∎

5.5 Spectrum Asymptotics for the Approximate Problems

In this section we prove an analog of Theorem 5.10 for the two approximate problems of Dirichlet half-droplet problem. Recall that set M is defined in (5.19).

Theorem 5.24. *(i) If $\{\lambda_l\}$, $l = 1, 2, ...$, is a sequence of eigenvalues to EVP (5.27) considered for either $i = 1$ or $i = 2$ corresponding to a sequence $\{\varepsilon_l\} \to 0$ and there exists a number $K^* > 0$ such that*

$$\left|\frac{\lambda_l}{\varepsilon_l^2}\right| \leq K^* \text{ for all } l \in \mathbb{N}, \tag{5.66}$$

then

$$\mathrm{dist}\left(\frac{\lambda_l}{\varepsilon_l^2}, M \setminus \{0\}\right) := \inf_{K \in M \setminus \{0\}} \left|K - \frac{\lambda_l}{\varepsilon_l^2}\right| \to 0.$$

(ii) Moreover, for sufficiently small $\varepsilon > 0$ and any eigenvalue λ of EVP (5.27) considered for either $i = 1$ or $i = 2$ one has

$$\lambda \notin \left(0,\, \left[\frac{\pi}{4(L - A/P)} \varepsilon\right]^2\right).$$

Proof: We show the proof only for the approximate problem "from below". In the exact same way one can show it for the approximate problem "from above". Let us fix $\tilde{\varepsilon} > 0$ and a family $[h_\varepsilon, \lambda_\varepsilon]$ of solutions to (5.27) with $i = 1$ for $\varepsilon \in (0, \tilde{\varepsilon})$. Then by assertion (i) of Proposition 5.19 and definitions (5.24a)–(5.24b) in the droplet core region one has

$$h^{(4)}_\varepsilon(x) + 2\varepsilon^3 h''_\varepsilon(x) = \varepsilon^{9/4} \lambda_\varepsilon h_\varepsilon(x) \quad \text{for} \quad x \in (0,\, a_\varepsilon/\varepsilon), \tag{5.67}$$

5.5 Spectrum Asymptotics for the Approximate Problems

Let us denote by $\phi_i(x, \lambda_\varepsilon)$ a fundamental system to (5.67) such that
$$\phi_i^{(k)}(0, \lambda_\varepsilon) = \varepsilon^k \delta_{i,k};\ i, k = 0, 1, 2, 3.$$
One can easily deduce that
$$\phi_1(x, \lambda_\varepsilon) = \varepsilon \left(\frac{z_{1,-}^2 \sinh(z_{1,+} x)}{z_{1,+}(z_{1,-}^2 + z_{1,+}^2)} + \frac{z_{1,+}^2 \sin(z_{1,-} x)}{z_{1,-}(z_{1,-}^2 + z_{1,+}^2)} \right),$$
$$\phi_3(x, \lambda_\varepsilon) = \varepsilon^3 \left(\frac{\sinh(z_{1,+} x)}{z_{1,+}(z_{1,-}^2 + z_{1,+}^2)} - \frac{\sin(z_{1,-} x)}{z_{1,-}(z_{1,-}^2 + z_{1,+}^2)} \right),$$
where we denote
$$z_{1,-} := \frac{1}{2}\sqrt{4\varepsilon^3 + 2\sqrt{4\varepsilon^6 + 4\lambda_\varepsilon \varepsilon^{9/4}}},\ z_{1,+} := \frac{1}{2}\sqrt{-4\varepsilon^3 + 2\sqrt{4\varepsilon^6 + 4\lambda_\varepsilon \varepsilon^{9/4}}}.$$

By assertion (iv) of Proposition 5.19 $h_\varepsilon(0) = h_\varepsilon''(0) = 0$, and therefore
$$h_\varepsilon(x) = C_\varepsilon^1 \phi_1(x, \lambda_\varepsilon) + C_\varepsilon^2 \phi_3(x, \lambda_\varepsilon) \quad \text{for} \quad x \in (0, a_\varepsilon/\varepsilon), \tag{5.68}$$
where C_ε^p, $p = 1, 2$ do not depend on x. By Proposition 5.19 and definitions (5.24a)–(5.24b) in the contact line region one has
$$h_\varepsilon^{(4)}(x) + k_1 h_\varepsilon''(x) = \lambda_\varepsilon h_\varepsilon(x) \quad \text{for} \quad x \in (a_\varepsilon/\varepsilon, b_\varepsilon/\varepsilon).$$
If one denotes by $\psi_i(x, \lambda_\varepsilon)$ a fundamental system to the last equation such that
$$\psi_i^{(k)}(a_\varepsilon/\varepsilon, \lambda_\varepsilon) = \delta_{i,k};\ i, k = 0, 1, 2, 3,$$
then one has
$$h_\varepsilon(x) = \sum_{i=0}^{3} h_\varepsilon^{(i)}(a_\varepsilon/\varepsilon + 0) \psi_i(x, \lambda_\varepsilon) \quad \text{for} \quad x \in (a_\varepsilon/\varepsilon, b_\varepsilon/\varepsilon). \tag{5.69}$$
It is easy to check that
$$\psi_0(x, \lambda_\varepsilon) = \frac{z_{2,-}^2 \cosh(z_{2,+}(x - a_\varepsilon/\varepsilon))}{(z_{2,-}^2 + z_{2,+}^2)} + \frac{z_{2,+}^2 \cos(z_{2,-}(x - a_\varepsilon/\varepsilon))}{(z_{2,-}^2 + z_{2,+}^2)},$$
$$\psi_1(x, \lambda_\varepsilon) = \frac{z_{2,-}^2 \sinh(z_{2,+}(x - a_\varepsilon/\varepsilon))}{z_{2,+}(z_{2,-}^2 + z_{2,+}^2)} + \frac{z_{2,+}^2 \sin(z_{2,-}(x - a_\varepsilon/\varepsilon))}{z_{2,-}(z_{2,-}^2 + z_{2,+}^2)},$$
$$\psi_2(x, \lambda_\varepsilon) = \frac{\cosh(z_{2,+}(x - a_\varepsilon/\varepsilon))}{(z_{2,-}^2 + z_{2,+}^2)} - \frac{\cos(z_{2,-}(x - a_\varepsilon/\varepsilon))}{(z_{2,-}^2 + z_{2,+}^2)},$$
$$\psi_3(x, \lambda_\varepsilon) = \frac{\sinh(z_{2,+}(x - a_\varepsilon/\varepsilon))}{z_{2,+}(z_{2,-}^2 + z_{2,+}^2)} - \frac{\sin(z_{2,-}(x - a_\varepsilon/\varepsilon))}{z_{2,-}(z_{2,-}^2 + z_{2,+}^2)},$$
where we denote
$$z_{2,-} := \frac{1}{2}\sqrt{2k_1 + 2\sqrt{k_1^2 + 4\lambda_\varepsilon}},\ z_{2,+} := \frac{1}{2}\sqrt{-2k_1 + 2\sqrt{k_1^2 + 4\lambda_\varepsilon}}.$$
Finally, in the outer interval $[h_\varepsilon, \lambda_\varepsilon]$ satisfies
$$h_\varepsilon^{(4)}(x) - (1 - \varepsilon^{1/12}) h_\varepsilon''(x) = \lambda_\varepsilon h_\varepsilon(x) \quad \text{for} \quad x \in (b_\varepsilon/\varepsilon, L/\varepsilon).$$

Chapter 5 Spectrum Asymptotics in a Singular Limit

Using this and $h_\varepsilon(L/\varepsilon) = h_\varepsilon''(L/\varepsilon) = 0$ one can write

$$h_\varepsilon(x) = -C_\varepsilon^3 \sin\left(z_{3,-}(x - L/\varepsilon)\right) - C_\varepsilon^4 \sinh\left(z_{3,+}(x - L/\varepsilon)\right) \text{ for } x \in (b_\varepsilon/\varepsilon,\ L/\varepsilon), \tag{5.70}$$

where C_ε^p, $p = 3, 4$ do not depend on x and

$$z_{3,-} := \frac{1}{2}\sqrt{-2(1 - \varepsilon^{1/12}) + 2\sqrt{(1 - \varepsilon^{1/12})^2 + 4\lambda_\varepsilon}},$$

$$z_{3,+} := \frac{1}{2}\sqrt{2(1 - \varepsilon^{1/12}) + 2\sqrt{(1 - \varepsilon^{1/12})^2 + 4\lambda_\varepsilon}}.$$

Let us denote $\psi_{i,b_\varepsilon}^{(k)} := \psi_i^{(k)}(b_\varepsilon/\varepsilon, \lambda_\varepsilon)$ and $\phi_{i,a_\varepsilon}^{(k)} := \phi_i^{(k)}(a_\varepsilon/\varepsilon, \lambda_\varepsilon)$ for $i, k = 0, 1, 2, 3$. Then using representations (5.68), (5.69), (5.70) as well as connection conditions (5.29), (5.30b) for $h_\varepsilon(x)$ at the points $x = a_\varepsilon/\varepsilon$ and $x = b_\varepsilon/\varepsilon$ one can construct a system of linear algebraic equations imposed for each $\varepsilon \in (0, \tilde{\varepsilon})$ on C_ε^p, $p = 1...4$ in the following form:

$$\begin{pmatrix} \gamma_{1,1} & \gamma_{1,2} & \sin(z_{3,-}(a_\varepsilon - L)/\varepsilon) & \sinh(z_{3,+}(a_\varepsilon - L)/\varepsilon) \\ \gamma_{2,1} & \gamma_{2,2} & z_{3,-}\cos(z_{3,-}(a_\varepsilon - L)/\varepsilon) & z_{3,+}\cosh(z_{3,+}(a_\varepsilon - L)/\varepsilon) \\ \gamma_{3,1} & \gamma_{3,2} & -z_{3,-}^2\sin(z_{3,-}(a_\varepsilon - L)/\varepsilon) & z_{3,+}^2\sinh(z_{3,+}(a_\varepsilon - L)/\varepsilon) \\ \gamma_{4,1} & \gamma_{4,2} & -z_{3,-}^3\cos(z_{3,-}(a_\varepsilon - L)/\varepsilon) & z_{3,+}^3\cosh(z_{3,+}(a_\varepsilon - L)/\varepsilon) \end{pmatrix} \begin{bmatrix} C_\varepsilon^1 \\ C_\varepsilon^2 \\ C_\varepsilon^3 \\ C_\varepsilon^4 \end{bmatrix} = 0, \tag{5.71}$$

where we denoted

$$\Gamma_\varepsilon = \begin{pmatrix} \gamma_{1,1} & \gamma_{1,2} \\ \gamma_{2,1} & \gamma_{2,2} \\ \gamma_{3,1} & \gamma_{3,2} \\ \gamma_{4,1} & \gamma_{4,2} \end{pmatrix} := \Psi_\varepsilon \cdot \Phi_\varepsilon, \tag{5.72a}$$

$$\Psi_\varepsilon := \begin{pmatrix} \psi_{0,b_\varepsilon}^{(0)} & \psi_{1,b_\varepsilon}^{(0)} & \psi_{2,b_\varepsilon}^{(0)} & \psi_{3,b_\varepsilon}^{(0)} \\ \psi_{0,b_\varepsilon}^{(1)} & \psi_{1,b_\varepsilon}^{(1)} & \psi_{2,b_\varepsilon}^{(1)} & \psi_{3,b_\varepsilon}^{(1)} \\ \psi_{0,b_\varepsilon}^{(2)} & \psi_{1,b_\varepsilon}^{(2)} & \psi_{2,b_\varepsilon}^{(2)} & \psi_{3,b_\varepsilon}^{(2)} \\ \psi_{0,b_\varepsilon}^{(3)} + k\psi_{0,b_\varepsilon}^{(1)} & \psi_{1,b_\varepsilon}^{(3)} + k\psi_{1,b_\varepsilon}^{(1)} & \psi_{2,b_\varepsilon}^{(3)} + k\psi_{2,b_\varepsilon}^{(1)} & \psi_{3,b_\varepsilon}^{(3)} + k\psi_{3,b_\varepsilon}^{(1)} \end{pmatrix}, \tag{5.72b}$$

$$\Phi_\varepsilon := \begin{pmatrix} \phi_{1,a_\varepsilon}^{(0)} & \phi_{3,a_\varepsilon}^{(0)} \\ \phi_{1,a_\varepsilon}^{(1)} & \phi_{3,a_\varepsilon}^{(1)} \\ \phi_{1,a_\varepsilon}^{(2)} & \phi_{3,a_\varepsilon}^{(2)} \\ \phi_{1,a_\varepsilon}^{(3)} - k_1\phi_{1,a_\varepsilon}^{(1)} & \phi_{3,a_\varepsilon}^{(3)} - k_1\phi_{3,a_\varepsilon}^{(1)} \end{pmatrix}, \tag{5.72c}$$

and $k = k_1 + 1 - \varepsilon^{1/12}$ is defined in (5.29). The homogeneous linear system of equations (5.71) has a nontrivial solution for each $\varepsilon \in (0, \tilde{\varepsilon})$ if and only if its determinant is zero identically in ε. Expanding its determinant in the third column this implies

$$0 \equiv \sin\left(z_{3,-}(a_\varepsilon - L)/\varepsilon\right) N_\varepsilon - z_{3,-}\cos\left(z_{3,-}(a_\varepsilon - L)/\varepsilon\right) D_\varepsilon,$$

5.5 Spectrum Asymptotics for the Approximate Problems

where we denoted two minors as

$$N_\varepsilon := \begin{vmatrix} \gamma_{2,1} & \gamma_{2,2} & z_{3,+}\cosh\left(z_{3,+}(a_\varepsilon - L)/\varepsilon\right) \\ \gamma_{3,1} + z_{3,-}^2\gamma_{1,1} & \gamma_{3,2} + z_{3,-}^2\gamma_{1,2} & \left(z_{3,+}^2 + z_{3,-}^2\right)\sinh\left(z_{3,+}(a_\varepsilon - L)/\varepsilon\right) \\ \gamma_{4,1} + z_{3,-}^2\gamma_{2,1} & \gamma_{4,2} + z_{3,-}^2\gamma_{2,2} & z_{3,+}\left(z_{3,+}^2 + z_{3,-}^2\right)\cosh\left(z_{3,+}(a_\varepsilon - L)/\varepsilon\right) \end{vmatrix},$$

$$D_\varepsilon := \begin{vmatrix} \gamma_{1,1} & \gamma_{1,2} & \sinh\left(z_{3,+}(a_\varepsilon - L)/\varepsilon\right) \\ \gamma_{3,1} + z_{3,-}^2\gamma_{1,1} & \gamma_{3,2} + z_{3,-}^2\gamma_{1,2} & \left(z_{3,+}^2 + z_{3,-}^2\right)\sinh\left(z_{3,+}(a_\varepsilon - L)/\varepsilon\right) \\ \gamma_{4,1} + z_{3,-}^2\gamma_{2,1} & \gamma_{4,2} + z_{3,-}^2\gamma_{2,2} & z_{3,+}\left(z_{3,+}^2 + z_{3,-}^2\right)\cosh\left(z_{3,+}(a_\varepsilon - L)/\varepsilon\right) \end{vmatrix}.$$

Therefore, one obtains

$$\cot\left(z_{3,-}(a_\varepsilon - L)/\varepsilon\right) \equiv \frac{N_\varepsilon}{z_{3,-}D_\varepsilon} \tag{5.73}$$

Next, we denote $K_\varepsilon := \lambda_\varepsilon/\varepsilon^2$. Let us first describe the case $K^* \geq K_\varepsilon > 0$ for all $\varepsilon \in (0, \tilde{\varepsilon})$, where constant K^* does not depend on ε. Using this and definition of $z_{3,\pm}$ one obtains

$$z_{3,-} \sim \sqrt{K_\varepsilon}\varepsilon \quad \text{and} \quad z_{3,+} \sim 1. \tag{5.74}$$

Using this and the assertion (ii) of Lemma 5.8 one obtains

$$\cot\left(z_{3,-}(a_\varepsilon - L)/\varepsilon\right) \sim \cot\left(\sqrt{K_\varepsilon}(a_\varepsilon - L)\right)$$
$$\coth\left(z_{3,+}(a_\varepsilon - L)/\varepsilon\right) \sim \tanh\left(z_{3,+}(a_\varepsilon - L)/\varepsilon\right) \sim 1. \tag{5.75}$$

Applying last three asymptotics to (5.73) results in

$$\cot\left(\sqrt{K_\varepsilon}(a_\varepsilon - L)\right) \sim \frac{\begin{vmatrix} \gamma_{2,1} & \gamma_{2,2} & 1 \\ \gamma_{3,1} + K_\varepsilon\varepsilon^2\gamma_{1,1} & \gamma_{3,2} + K_\varepsilon\varepsilon^2\gamma_{1,2} & 1 \\ \gamma_{4,1} + K_\varepsilon\varepsilon^2\gamma_{2,1} & \gamma_{4,2} + K_\varepsilon\varepsilon^2\gamma_{2,2} & 1 \end{vmatrix}}{\sqrt{K_\varepsilon}\varepsilon \begin{vmatrix} \gamma_{1,1} & \gamma_{1,2} & 1 \\ \gamma_{3,1} + K_\varepsilon\varepsilon^2\gamma_{1,1} & \gamma_{3,2} + K_\varepsilon\varepsilon^2\gamma_{1,2} & 1 \\ \gamma_{4,1} + K_\varepsilon\varepsilon^2\gamma_{2,1} & \gamma_{4,2} + K_\varepsilon\varepsilon^2\gamma_{2,2} & 1 \end{vmatrix}}. \tag{5.76}$$

Let us now derive the asymptotics for matrix Γ_ε as $\varepsilon \to 0$. Below we apply symbol $'\sim'$ for matrices to denote their element-wise asymptotic equivalence in the sense of Definition 1.1. Analogously to (5.74) by definitions of $z_{1,\pm}$, $z_{2,\pm}$ one obtains

$$z_{1,-} \sim z_{1,+} \sim K_\varepsilon^{1/4}\varepsilon \quad \text{and} \quad z_{2,-} \sim \sqrt{k_1}, \quad z_{2,+} \sim \sqrt{\frac{K_\varepsilon}{k_1}}\varepsilon.$$

This and definition of ϕ_i, $i = 1, 3$ imply that for all $x \in (0, a_\varepsilon/\varepsilon)$

$$\phi_1(x, \lambda_\varepsilon) \sim \varepsilon x, \qquad \phi_3(x, \lambda_\varepsilon) \sim \varepsilon^3 x^3/6;$$
$$\phi_1'(x, \lambda_\varepsilon) \sim \varepsilon, \qquad \phi_3'(x, \lambda_\varepsilon) \sim \varepsilon^3 x^2/2;$$
$$\phi_1''(x, \lambda_\varepsilon) \sim \varepsilon^6 \text{const} x^3/6, \quad \phi_3''(x, \lambda_\varepsilon) \sim \varepsilon^3 x;$$
$$\phi_1'''(x, \lambda_\varepsilon) \sim \varepsilon^5 K_\varepsilon, \qquad \phi_3'''(x, \lambda_\varepsilon) \sim \varepsilon^3. \tag{5.77}$$

Therefore, by definition (5.72c) one has

$$\Phi_\varepsilon \sim \begin{pmatrix} a_\varepsilon & \frac{a_\varepsilon^3}{6} \\ \varepsilon & \varepsilon \frac{a_\varepsilon^2}{2} \\ \varepsilon^3 \text{const} & \varepsilon^2 a_\varepsilon \\ -\varepsilon k_1 & -\varepsilon k_1 \frac{a_\varepsilon^2}{2} \end{pmatrix}$$

Similarly using definition (5.72b) and assertion (ii) of Lemma 5.8 one obtains

$$\Psi_\varepsilon \sim \begin{pmatrix} 1 & \frac{\sin \rho_\varepsilon}{\sqrt{k_1}} & \frac{1-\cos \rho_\varepsilon}{k_1} & \frac{\sqrt{K_\varepsilon}\rho_\varepsilon}{k_1^{5/2}}\varepsilon \\ \frac{K_\varepsilon \rho_\varepsilon}{k_1^{3/2}}\varepsilon^2 & \cos \rho_\varepsilon & \frac{\sin \rho_\varepsilon}{\sqrt{k_1}} & \frac{\sqrt{K_\varepsilon}(1-\cos \rho_\varepsilon)}{k_1^2}\varepsilon \\ \frac{K_\varepsilon(1+\cos \rho_\varepsilon)}{k_1}\varepsilon^2 & -\sqrt{k_1}\sin \rho_\varepsilon & -\cos \rho_\varepsilon & \frac{\sqrt{K_\varepsilon}\sin \rho_\varepsilon}{k_1^{3/2}}\varepsilon \\ \frac{K_\varepsilon \rho_\varepsilon(1+k_1)}{k_1^{3/2}}\varepsilon^2 & \cos \rho_\varepsilon & \frac{\sin \rho_\varepsilon}{\sqrt{k_1}} & \frac{\sqrt{K_\varepsilon}(1+k_1-\cos \rho_\varepsilon)}{k_1}\varepsilon \end{pmatrix},$$

where we denoted

$$\rho_\varepsilon := \sqrt{k_1}(b_\varepsilon - a_\varepsilon)/\varepsilon. \tag{5.78}$$

Using the simple rule

$$f_1(\varepsilon) \sim f_2(\varepsilon),\ g_1(\varepsilon) \sim g_2(\varepsilon) \implies f_1(\varepsilon) \cdot g_1(\varepsilon) \sim f_2(\varepsilon) \cdot g_2(\varepsilon),$$

definition (5.72a) together with asymptotics for Φ_ε, Ψ_ε and the fact $b_\varepsilon - a_\varepsilon = O(\varepsilon^{1/4})$ one obtains

$$\Gamma_\varepsilon \sim \begin{pmatrix} a_\varepsilon & \frac{a_\varepsilon^3}{6} \\ \varepsilon \cos \rho_\varepsilon & \varepsilon \frac{a_\varepsilon^2}{2} \cos \rho_\varepsilon \\ -\varepsilon \sqrt{k_1} \sin \rho_\varepsilon & -\varepsilon \sqrt{k_1} \frac{a_\varepsilon^2}{2} \sin \rho_\varepsilon \\ \varepsilon \cos \rho_\varepsilon & \varepsilon \frac{a_\varepsilon^2}{2} \cos \rho_\varepsilon \end{pmatrix}. \tag{5.79}$$

Finally, from this and (5.72a), (5.76) one gets:

$$\cot\left(\sqrt{K_\varepsilon}(a_\varepsilon - L)\right) \sim \frac{O(\varepsilon)}{\sqrt{K_\varepsilon}(\cos \rho_\varepsilon + \sqrt{k_1}\sin \rho_\varepsilon)}. \tag{5.80}$$

The last asymptotics prohibits sequences $\{\varepsilon_l\} \to 0$ and $\{K_{\varepsilon_l}\}$ such that $K_{\varepsilon_l} \to 0$ and $K^* > K_{\varepsilon_l} > 0$ for all $l \in \mathbb{N}$, because in such case $\cot \sqrt{K_\varepsilon}(a_{\varepsilon_l} - L) \sim 1/\sqrt{K_{\varepsilon_l}}$ and this would contradict to (5.80). Therefore, without loss of generality (see also Remark 5.26 below) one obtains that $\cot \sqrt{K_\varepsilon}(a_\varepsilon - L) \to 0$ as $\varepsilon \to 0$. From this one concludes that $K_\varepsilon \to M \setminus \{0\}$, where set M is defined in (5.19).

Next, we consider the case $-K^* \leq K_\varepsilon < 0$ for all $\varepsilon \in (0, \tilde{\varepsilon})$ and substitute it again in expression (5.73). As before after derivation of the leading order asymptotics for (5.73) for this case one obtains that asymptotic balance (5.80) transforms to

$$\coth\left(\sqrt{-K_\varepsilon}(a_\varepsilon - L)\right) \sim \frac{O(\varepsilon)}{\sqrt{-K_\varepsilon}(\cos \rho_\varepsilon + \sqrt{k_1}\sin \rho_\varepsilon)}. \tag{5.81}$$

Again firstly we obtain from it that sequences $\{\varepsilon_l\} \to 0$ and $\{K_{\varepsilon_l}\}$ such that $K_{\varepsilon_l} \to 0$ and $-K^* < K_{\varepsilon_l} < 0$ for all $l \in \mathbb{N}$ are not possible. But then right-hand side of (5.81) tends to zero as $\varepsilon \to 0$ and we arrive to a contradiction because function $\coth\left(\sqrt{-K_\varepsilon}(a_\varepsilon - L)\right)$ is bounded away from 0. Therefore, the case $-K^* \leq K_\varepsilon < 0$ for all $\varepsilon \in (0, \tilde{\varepsilon})$ is not possible. Proceeding similarly one can show that the case $K_\varepsilon \equiv 0$ for all $\varepsilon \in (0, \tilde{\varepsilon})$ is not possible as well.

We conclude that if there exists a constant $K^* > 0$ such that $|K_\varepsilon| \leq K^*$ for all $\varepsilon \in (0, \tilde{\varepsilon})$ then one has $K_\varepsilon > 0$ for sufficiently small $\varepsilon > 0$ and $K_\varepsilon \to M \setminus \{0\}$ as $\varepsilon \to 0$. This in fact implies both assertions of the theorem. ∎

Remark 5.25. Let us point out the difference between Dirichlet (5.21) and Neumann (5.22) half-droplet problems. Analogously to the proof of Theorem 5.24 one can show that all assertions of Theorem 5.10 hold for the approximate problems for Neumann half-droplet problem (5.18). The difference between approximate Neumann and Dirichlet half-droplet problems lies in the fact that for the former ones there may exists sequences of eigenvalues $\{\lambda_l\}$ and $\{\varepsilon_l\} \to 0$ that satisfy condition (5.66) and have

$$\frac{\lambda_l}{\varepsilon_l^2} \to 0 \text{ as } l \to \infty.$$

Indeed, proceeding analogously to the proof above one obtains that analogs of asymptotic balances (5.80)–(5.81) in the case of approximate Neumann half-droplet problems are

$$\cot\left(\sqrt{K_\varepsilon}(a_\varepsilon - L)\right) \sim \frac{O(\varepsilon)}{\sqrt{K_\varepsilon^3}(\cos\rho_\varepsilon + \sqrt{k_1}\sin\rho_\varepsilon)}$$

and

$$\coth\left(\sqrt{-K_\varepsilon}(a_\varepsilon - L)\right) \sim \frac{O(\varepsilon)}{\sqrt{(-K_\varepsilon)^3}(\cos\rho_\varepsilon + \sqrt{k_1}\sin\rho_\varepsilon)},$$

respectively. Analyzing them one can easily check that in this case there could exists sequences $\{\varepsilon_l\} \to 0$ and $\{K_{\varepsilon_l}\}$ such that $K_{\varepsilon_l} \to 0$ as $l \to \infty$. This result stays in agreement with Theorem 5.12 which asserts that symmetric EVP (5.18) possesses one exponentially small eigenvalue as $\varepsilon \to 0$, which turns out to satisfy Neumann half-droplet problem (5.22). ∎

Remark 5.26. One can wonder what happens for asymptotics (5.80)–(5.81) if one takes a sequence $\{\varepsilon_l\} \to 0$, such that $\sin(\rho_{\varepsilon_l} + \varphi) \to 0$ as $l \to \infty$, where

$$\varphi := \arcsin 1/\sqrt{1+k_1} \in (0, \pi/2).$$

Clearly, in this case it may happen that $\cot\sqrt{K_{\varepsilon_l}}(a_{\varepsilon_l} - L)$ does not tend to zero and the assertions of Theorem 5.24 become unclear. To escape from this situation one should recall definition (5.78) and use that there exists a certain freedom in defining functions a_ε and b_ε with properties stated in Lemma 5.8. One can redefine the contact line region $(a_\varepsilon/\varepsilon, b_\varepsilon/\varepsilon)$ for all ε belonging to a special set $\mathcal{O} \subset \mathbb{R}$ so that for any sequence $\{\varepsilon_l\} \to 0$ one would have

$$|\sin(\sqrt{k_1}(b_{\varepsilon_l} - a_{\varepsilon_l})/\varepsilon_l + \varphi)| \geq \text{const} > 0 \text{ for all } l \in \mathbb{N}. \tag{5.82}$$

and all the assertions of Lemmata 5.8, 5.17 would hold with redefined a_ε, b_ε as well without any changes for results of this chapter. Namely, let $a_\varepsilon = a(\varepsilon)$ and $b_\varepsilon = b(\varepsilon)$ satisfy assertions of Lemma 5.8. Set $\mathcal{O}$ can be defined for example as follows:

$$\mathcal{O} = \left\{\varepsilon > 0 : \exists n \in \mathbb{N},\ \frac{\sqrt{k_1}(b(\varepsilon) - a(\varepsilon))}{\varepsilon} \in (-7/6\varphi + \pi n, -5/6\varphi + \pi n)\right\}. \tag{5.83}$$

Chapter 5 Spectrum Asymptotics in a Singular Limit

Next, redefine functions $a_\varepsilon, b_\varepsilon$ as

$$a_\varepsilon := \begin{cases} a(\varepsilon), & \varepsilon \notin \mathcal{O} \\ a(\varepsilon) - \dfrac{b(\varepsilon) - a(\varepsilon)}{2}, & \varepsilon \in \mathcal{O} \end{cases}, \quad b_\varepsilon := \begin{cases} b(\varepsilon), & \varepsilon \notin \mathcal{O} \\ b(\varepsilon) + \dfrac{b(\varepsilon) - a(\varepsilon)}{2}, & \varepsilon \in \mathcal{O} \end{cases} \quad (5.84)$$

and fix any sequence $\{\varepsilon_l\} \to 0$. It can be decomposed into two subsequences $\{\varepsilon_{l_k}\}, \{\varepsilon_{l_m}\} \to 0$ (one of which may be empty or finite) such that $\varepsilon_{l_k} \in \mathcal{O}$ for all $k \in \mathbb{N}$ and $\varepsilon_{l_m} \notin \mathcal{O}$ for all $m \in \mathbb{N}$. Then by definitions (5.83) and (5.84) one obtains

$$|\sin(\sqrt{k_1}(b_{\varepsilon_{l_k}} - a_{\varepsilon_{l_k}})/\varepsilon_{l_k} + \varphi)| = |\sin(2\sqrt{k_1}(b(\varepsilon_{l_k}) - a(\varepsilon_{l_k}))/\varepsilon_{l_k} + \varphi)| \geq \sin(2/3\varphi) > 0,$$
$$|\sin(\sqrt{k_1}(b_{\varepsilon_{l_m}} - a_{\varepsilon_{l_m}})/\varepsilon_{l_m} + \varphi)| = |\sin(\sqrt{k_1}(b(\varepsilon_{l_m}) - a(\varepsilon_{l_m}))/\varepsilon_{l_m} + \varphi)| \geq \sin(1/6\varphi) > 0$$

for all $k, m \in \mathbb{N}$. We conclude that new definition (5.84) makes all assertions of Lemmata 5.8, 5.17 to be fulfilled again and (5.82) holds for any sequence $\{\varepsilon_l\} \to 0$. Therefore, (5.80)–(5.81) really imply all assertions of Theorem 5.24. ∎

At the end of this section we prove a lemma which we use in section 5.7 for the proof of uniqueness assertion (iii) of Theorem 5.11. Besides it gives leading orders as $\varepsilon \to 0$ for eigenfunctions of approximate problems.

Lemma 5.27. *Let numbers $k, m \in \mathbb{N}_0$ and corresponding $\lambda_\varepsilon^{i,k}$ and $\lambda_\varepsilon^{i,m}$ be k-th and m-th eigenvalues, respectively, from the ordering (5.28) for the approximate eigenvalue problem (5.27) with fixed $i = 1$ or $i = 2$. If there exists a number $K^* > 0$ such that*

$$\left| \frac{\lambda_\varepsilon^{i,k}}{\varepsilon^2} \right| \leq K^* \quad \text{and} \quad \left| \frac{\lambda_\varepsilon^{i,m}}{\varepsilon^2} \right| \leq K^*$$

for all sufficiently small $\varepsilon > 0$ then there exist positive numbers K^{**} and $\tilde{\varepsilon}$ such that

$$\left| \lambda_\varepsilon^{i,k} - \lambda_\varepsilon^{i,m} \right| \geq K^{**} \varepsilon^2$$

for all $\varepsilon \in (0, \tilde{\varepsilon})$.

Proof: Let us prove the lemma using a contradiction argument. We do it again only for the case (5.27) with $i = 1$. For the case $i = 2$ the proof is analogous. Suppose that the assertion of the lemma is not true. Then by assertion (i) of Theorem 5.24 it follows that there exist a positive number $K \in M \setminus \{0\}$ and sequences $\{\varepsilon_l\} \to 0, \{\lambda_{\varepsilon_l}\}, \{\tilde{\lambda}_{\varepsilon_l}\}$ such that

$$\lambda_{\varepsilon_l} := \lambda_{\varepsilon_l}^{1,k}, \quad \tilde{\lambda}_{\varepsilon_l} := \lambda_{\varepsilon_l}^{1,m} \text{ for each } l \in \mathbb{N},$$
$$\frac{\lambda_{\varepsilon_l}}{\varepsilon_l^2} \to K, \quad \frac{\tilde{\lambda}_{\varepsilon_l}}{\varepsilon_l^2} \to K \text{ as } l \to \infty. \quad (5.85)$$

Let $\{h_{\varepsilon_l}\}$ be the sequence of eigenfunctions h_{ε_l} corresponding to λ_{ε_l} for each $l \in \mathbb{N}$. Following the lines of the proof for Theorem 5.24 one obtains that there exist $C_{\varepsilon_l}^p$, $p = 1, 2, 3, 4$ such that representations (5.68), (5.69), (5.70) hold for h_{ε_l} on the droplet core, contact line and outer layer intervals, respectively. Moreover, $C_{\varepsilon_l}^p$ are solutions of the homogeneous linear system (5.71) with the notation defined in the proof of Theorem 5.24. Let

$$\mathbb{M}_{\varepsilon_l} := \begin{pmatrix} \gamma_{1,2} & \sin(z_{3,-}(a_{\varepsilon_l} - L)/\varepsilon_l) & \sinh(z_{3,+}(a_{\varepsilon_l} - L)/\varepsilon_l) \\ \gamma_{2,2} & z_{3,-} \cos(z_{3,-}(a_{\varepsilon_l} - L)/\varepsilon_l) & z_{3,+} \cosh(z_{3,+}(a_{\varepsilon_l} - L)/\varepsilon_l) \\ \gamma_{3,2} & -z_{3,-}^2 \sin(z_{3,-}(a_{\varepsilon_l} - L)/\varepsilon_l) & z_{3,+}^2 \sinh(z_{3,+}(a_{\varepsilon_l} - L)/\varepsilon_l) \end{pmatrix}.$$

5.5 Spectrum Asymptotics for the Approximate Problems

We claim that there exists a subsequence $\{\lambda_{\varepsilon_l}\}$ (to avoid technicalities we denote all subsequences below also as $\{\lambda_{\varepsilon_l}\}$) such that $\det \mathbb{M}_{\varepsilon_l} \neq 0$ for all $l \in \mathbb{N}$. Suppose inverse then one can fix a subsequence $\{\lambda_{\varepsilon_l}\}$ such that $\det \mathbb{M}_{\varepsilon_l} \equiv 0$ for all $l \in \mathbb{N}$. Expanding the determinant of $\mathbb{M}_{\varepsilon_l}$ in the second column and dividing the resulting expression by $\sin(z_{3,-}(a_{\varepsilon_l} - L)/\varepsilon_l)$ one obtains:

$$\cot(z_{3,-}(a_{\varepsilon_l}-L)/\varepsilon_l) = \frac{\begin{vmatrix} \gamma_{2,2} & z_{3,+}\cosh(z_{3,+}(a_{\varepsilon_l}-L)/\varepsilon_l) \\ \gamma_{3,2} + z_{3,-}^2 \gamma_{1,2} & \left(z_{3,+}^2 + z_{3,-}^2\right)\sinh(z_{3,+}(a_{\varepsilon_l}-L)/\varepsilon_l) \end{vmatrix}}{z_{3,-}\begin{vmatrix} \gamma_{1,2} & \sinh(z_{3,+}(a_{\varepsilon_l}-L)/\varepsilon_l) \\ \gamma_{3,2} + z_{3,-}^2 \gamma_{1,2} & \left(z_{3,+}^2 + z_{3,-}^2\right)\sinh(z_{3,+}(a_{\varepsilon_l}-L)/\varepsilon_l) \end{vmatrix}}.$$

One can check that asymptotics (5.74), (5.75) and (5.79) hold in the current case with $K_{\varepsilon_l} \to K$ as $l \to \infty$ by (5.85). From these asymptotics it follows that the right hand side of the last expression is $O(1)$ and the left hand side tends to zero as $l \to \infty$. This gives a contradiction, and therefore, indeed, we can fix a subsequence $\{\lambda_{\varepsilon_l}\}$ such that $\det \mathbb{M}_{\varepsilon_l} \neq 0$ for all $l \in \mathbb{N}$. For such a subsequence the matrix of the linear system of algebraic equations (5.71) has the rank equal 3. Therefore, using Cramer's rule and fixing $C^1_{\varepsilon_l}$ for each $l \in \mathbb{N}$ one gets uniquely $C^p_{\varepsilon_l}$, $p = 2, 3, 4$ as the solution of a linear system:

$$\begin{pmatrix} \gamma_{1,2} & \sin(z_{3,-}(a_{\varepsilon_l}-L)/\varepsilon_l) & \sinh(z_{3,+}(a_{\varepsilon_l}-L)/\varepsilon_l) \\ \gamma_{2,2} & z_{3,-}\cos(z_{3,-}(a_{\varepsilon_l}-L)/\varepsilon_l) & z_{3,+}\cosh(z_{3,+}(a_{\varepsilon_l}-L)/\varepsilon_l) \\ \gamma_{3,2} & -z_{3,-}^2\sin(z_{3,-}(a_{\varepsilon_l}-L)/\varepsilon_l) & z_{3,+}^2\sinh(z_{3,+}(a_{\varepsilon_l}-L)/\varepsilon_l) \end{pmatrix} \begin{bmatrix} C^2_{\varepsilon_l} \\ C^3_{\varepsilon_l} \\ C^4_{\varepsilon_l} \end{bmatrix} = -C^1_{\varepsilon_l}\begin{bmatrix} \gamma_{1,1} \\ \gamma_{2,1} \\ \gamma_{3,1} \end{bmatrix}.$$

This allows us to obtain asymptotics for $C^p_{\varepsilon_l}$, $p = 2, 3, 4$ as $l \to \infty$. For example, due to Cramer's rule

$$\frac{C^2_{\varepsilon_l}}{C^1_{\varepsilon_l}} = -\frac{\begin{vmatrix} \gamma_{1,1} & \sin(z_{3,-}(a_{\varepsilon_l}-L)/\varepsilon_l) & \sinh(z_{3,+}(a_{\varepsilon_l}-L)/\varepsilon_l) \\ \gamma_{2,1} & z_{3,-}\cos(z_{3,-}(a_{\varepsilon_l}-L)/\varepsilon_l) & z_{3,+}\cosh(z_{3,+}(a_{\varepsilon_l}-L)/\varepsilon_l) \\ \gamma_{3,1} & -z_{3,-}^2\sin(z_{3,-}(a_{\varepsilon_l}-L)/\varepsilon_l) & z_{3,+}^2\sinh(z_{3,+}(a_{\varepsilon_l}-L)/\varepsilon_l) \end{vmatrix}}{\begin{vmatrix} \gamma_{1,2} & \sin(z_{3,-}(a_{\varepsilon_l}-L)/\varepsilon_l) & \sinh(z_{3,+}(a_{\varepsilon_l}-L)/\varepsilon_l) \\ \gamma_{2,2} & z_{3,-}\cos(z_{3,-}(a_{\varepsilon_l}-L)/\varepsilon_l) & z_{3,+}\cosh(z_{3,+}(a_{\varepsilon_l}-L)/\varepsilon_l) \\ \gamma_{3,2} & -z_{3,-}^2\sin(z_{3,-}(a_{\varepsilon_l}-L)/\varepsilon_l) & z_{3,+}^2\sinh(z_{3,+}(a_{\varepsilon_l}-L)/\varepsilon_l) \end{vmatrix}}.$$

Expanding the nominator and the denominator of the last expression and again using asymptotics (5.74)–(5.75), (5.79) one gets that $C^2_{\varepsilon_l}/C^1_{\varepsilon_l} = -2/a_{\varepsilon_l}^2 + o(1)$. Analogously, one can obtain that $C^3_{\varepsilon_l}/C^1_{\varepsilon_l} = -2a_{\varepsilon_l}/3 + o(1)$ and $C^4_{\varepsilon_l}/C^1_{\varepsilon_l}$ is exponentially small, namely $C^4_{\varepsilon_l}/C^1_{\varepsilon_l} \sim o(\varepsilon_l)\exp((a_{\varepsilon_l}-L)/\varepsilon_l)$.

Next, let $\{\widetilde{h}_{\varepsilon_l}\}$ be the sequence of eigenfunctions $\widetilde{h}_{\varepsilon_l}$ corresponding to $\widetilde{\lambda}_{\varepsilon_l}$ from (5.85) for each $l \in \mathbb{N}$. By assertion (iii) of Proposition 5.18 one has

$$(h_{\varepsilon_l}, \widetilde{h}_{\varepsilon_l})_{H_{\varepsilon_l}} \equiv 0 \quad \text{for all} \quad l \in \mathbb{N}, \tag{5.86}$$

where we use inner product (5.26). Denote for each $l \in \mathbb{N}$ by $\widetilde{C}^p_{\varepsilon_l}$ for $p = 1, 2, 3, 4$ the solutions of the linear system corresponding to the eigenfunction $\widetilde{h}_{\varepsilon_l}$. We can fix $C^1_{\varepsilon_l} \equiv \widetilde{C}^1_{\varepsilon_l}$ for all $l \in \mathbb{N}$ so that (5.86) still holds. Then by considerations above and (5.85) it follows that $C^p_{\varepsilon_l} \sim \widetilde{C}^p_{\varepsilon_l}$ for $p = 2, 3, 4$ as $l \to \infty$. From this, again (5.85) and representations (5.68), (5.69), (5.70) for $h_{\varepsilon_l}(x)$ and $\widetilde{h}_{\varepsilon_l}(x)$ one gets

$$(h_{\varepsilon_l}, \widetilde{h}_{\varepsilon_l})_{H_{\varepsilon_l}} \to 1 \text{ as } l \to \infty.$$

But the last asymptotics contradicts to (5.86). Therefore, we arrive to a contradiction and the assertion of the lemma is true. ∎

Chapter 5 Spectrum Asymptotics in a Singular Limit

Remark 5.28. (**about leading orders for eigenfunctions**) Results of Theorem 5.24 and Lemma 5.27 allow us for constructing approximations for eigenvalues and eigenfunctions of approximate EVPs (5.27), which are helpful when one wants to show existence of solutions with prescribed asymptotics for these problems. Indeed, Theorem 5.24 allows for eigenvalues $\lambda_\varepsilon \sim K\varepsilon^2$, where $K \in M \setminus \{0\}$. Suppose such an eigenvalue exists for the EVP "from below". Then for the corresponding eigenfunction h_ε representations (5.68), (5.69), (5.70) should hold. Moreover, by the proof of Lemma 5.27 it follows that h_ε can be normalized so that

$$C_\varepsilon^1 \equiv 1, \quad C_\varepsilon^2 \sim -2/a_\varepsilon^2, \quad C_\varepsilon^3 \sim -2a_\varepsilon/3, \quad C_\varepsilon^4 \sim o(\varepsilon)\exp\left((a_\varepsilon - L)/\varepsilon\right) \tag{5.87}$$

From this, representation (5.68) and asymptotics (5.77) it follows that on the droplet core interval $(0, a_\varepsilon/\varepsilon)$ to the leading order $h_\varepsilon(x)$ is a linear combination of polynomials and does not depend on K. One can explain this fact looking at equation (5.67) for h_ε on the droplet core. The term $2\varepsilon^3 h_\varepsilon''(x) - \varepsilon^{9/4}\lambda_\varepsilon h_\varepsilon(x)$ is small enough, so that the leading orders for the fundamental system on this interval are given by the solutions of the equation $h_\varepsilon^{(4)}(x) = 0$, i.e. by polynomials. Such property of (5.67) in turn takes place because defining the approximate problems (5.27) we explored asymptotics derived in Lemma 5.17 for the coefficients $r_\varepsilon(x)$, $f_\varepsilon(x)$ of the symmetric EVP (5.18).

Taking next the leading order as $\varepsilon \to 0$ in representation (5.69) one can see that on the contact line region $(a_\varepsilon/\varepsilon, b_\varepsilon/\varepsilon)$ to the leading order $h_\varepsilon(x)$ is constant and its derivatives actually oscillate with a high frequency proportional to $(b_\varepsilon - a_\varepsilon)/\varepsilon$. Such oscillations can make problems to resolve numerically derivatives of eigenfunctions in this region (see details in section 5.8). Finally, on the outer layer $(b_\varepsilon/\varepsilon, L/\varepsilon)$ due to (5.70) and asymptotics (5.74), (5.87) holding with $K_\varepsilon \to K$ as $\varepsilon \to 0$ one obtains that

$$h_\varepsilon(x) \sim C_\varepsilon^3 \sin\left(\sqrt{K}(\varepsilon x - L)\right)$$

and essentially depends on K. If we consider instead of the approximate problem "from below" the one "from above" we end up with the same leading orders on the droplet core and the outer layer for the eigenfunctions h_ε corresponding to $\lambda_\varepsilon \sim K\varepsilon^2$ with $K \in M \setminus \{0\}$. This fact we explore in the next section. ∎

5.6 Existence of Eigenvalues with Prescribed Asymptotics

In this section we prove an analog of Theorem 5.11 for the approximate problem "from above".

Theorem 5.29. *For every $j \in \mathbb{N}_0$ there exists a positive constant ε^j and a smooth map $\lambda^j \in C^1((0, \varepsilon^j), \mathbb{R})$ such that for all $\varepsilon \in (0, \varepsilon^j)$ the following holds:*

(i) $\lambda^j(\varepsilon)$ is an eigenvalue of approximate EVP (5.27) with $i = 2$,

(ii) $\left|\lambda^j(\varepsilon) - \left(\frac{\pi(2j+1)}{2(L-A/P)}\varepsilon\right)^2\right| = o(\varepsilon^4)$.

For the proof of Theorem 5.29 we use a certain modification of a implicit function theorem developed in Recke and Omel'chenko [5] which has minimal assumptions concerning continuity with respect to the control parameter.

Theorem 5.30. *Let K_1, K_2, K_3, ν_1, ν_2 be positive numbers with $\nu_1 > \nu_2$ and for all $\varepsilon \in (0, \varepsilon_0)$ be given Banach spaces Y_ε and Z_ε and maps $F_\varepsilon \in C^1(Y_\varepsilon, Z_\varepsilon)$ such that*

$$\|F_\varepsilon(0)\| \leq K_1 \varepsilon^{\nu_1}, \tag{5.88}$$

$$\|F_\varepsilon'(u) - F_\varepsilon'(0)\| \leq K_2 \frac{\|u\|}{\varepsilon^{\nu_2}} \tag{5.89}$$

and operators $F'_\varepsilon(0)$ are invertible with

$$||F'_\varepsilon(0)^{-1}|| \leq K_3. \tag{5.90}$$

Then there exists $\varepsilon_1 \in (0, \varepsilon_0)$ and $\delta > 0$ such that for all $\varepsilon \in (0, \varepsilon_1)$ there exists exactly one $u = u_\varepsilon$ with $||u|| < \delta \varepsilon^{\nu_2}$ and $F_\varepsilon(u) = 0$. Moreover,

$$||u_\varepsilon|| \leq 2K_3 ||F_\varepsilon(0)||. \tag{5.91}$$

Proof: For $\varepsilon \in (0, \varepsilon_0)$ one has $F_\varepsilon(u) = 0$ if and only if

$$G_\varepsilon(u) := u - F'_\varepsilon(0)^{-1} F_\varepsilon(u) = u. \tag{5.92}$$

Moreover, for such ε and all $u, v \in Y_\varepsilon$ one has

$$G_\varepsilon(u) - G_\varepsilon(v) = \int_0^1 G'_\varepsilon(su + (1-s)v)(u-v)\,ds =$$
$$= F'_\varepsilon(0)^{-1} \int_0^1 (F'_\varepsilon(su + (1-s)v) - F'_\varepsilon(0))\,(v-u)\,ds.$$

Assumptions (5.89) and (5.90) imply that there exists $\varepsilon_1 \in (0, \varepsilon_0)$ and $\delta > 0$ such that for all $\varepsilon \in (0, \varepsilon_1)$

$$||G_\varepsilon(u) - G_\varepsilon(v)|| \leq \frac{1}{2}||u - v|| \text{ for all } u, v \in R_\varepsilon := \{w \in Y_\varepsilon : ||w|| \leq \delta \varepsilon^{\nu_2}\}.$$

Using this and (5.88), (5.90) for all $\varepsilon \in (0, \varepsilon_1)$ and $u \in R_\varepsilon$ one gets

$$||G_\varepsilon(u)|| \leq ||G_\varepsilon(u) - G_\varepsilon(0)|| + ||G_\varepsilon(0)|| \leq \frac{1}{2}||u|| + K_3||F_\varepsilon(0)|| \leq \frac{\delta}{2}\varepsilon^{\nu_2} + K_1 K_3 \varepsilon^{\nu_1}. \tag{5.93}$$

From this and condition $\nu_1 > \nu_2$ it follows that G_ε maps R_ε into R_ε for all $\varepsilon \in (0, \varepsilon_1)$, if one chooses ε_1 sufficiently small. Now, Banach's fixed point theorem gives a unique in R_ε solution $u = u_\varepsilon$ to (5.92) for all $\varepsilon \in (0, \varepsilon_1)$. Moreover, (5.93) yields

$$||u_\varepsilon|| \leq \frac{1}{2}||u_\varepsilon|| + K_3 ||F_\varepsilon(0)||,$$

i.e. (5.91). ∎

The next lemma (see Magnus [41] or Recke and Omel'chenko [5]) allows sometimes to prove assumption (5.90).

Lemma 5.31. Let $F'_\varepsilon(0)$ be Fredholm of index zero for all $\varepsilon \in (0, \varepsilon_0)$. Suppose that there do not exist sequences $\varepsilon_1, \varepsilon_2, ... \in (0, \varepsilon_0)$ and $u_1 \in U_{\varepsilon_1}, u_2 \in U_{\varepsilon_2}...$ with $||u_n|| = 1$ for all $n \in \mathbb{N}$ and $|\varepsilon_n| + ||F'_\varepsilon(u)u_n|| \to 0$ for $n \to 0$. Then assumption (5.90) is satisfied.

Proof of Theorem 5.29:
Step1–reformulation of EVP (5.18) with $i = 2$ into the framework of Theorem 5.30:
Define a bilinear form

$$g_\varepsilon(h, w) := \int_0^{L/\varepsilon} (h'' w'' - r_\varepsilon^2 h' w')\,dx \text{ for } h, w \in H^2(0, L/\varepsilon) \cap H_0^1(0, L/\varepsilon). \tag{5.94}$$

Chapter 5 Spectrum Asymptotics in a Singular Limit

EVP problem "from above" can be reformulated as

$$h \in V_\varepsilon, \lambda \in \mathbb{R} : g_\varepsilon(h, w) = \lambda (h, w)_{H_\varepsilon}, \ \forall w \in V_\varepsilon,$$

where $(h, w)_{H_\varepsilon}$ is the inner product (5.26) in Hilbert space H_ε defined in (5.25) and Hilbert space V_ε is defined in (5.23). Recall that V_ε is equipped with the inner product

$$(h, w)_{V_\varepsilon} := \sum_{k=0}^{2} \int_{0}^{L/\varepsilon} h^{(k)}(x) w^{(k)}(x) \, dx.$$

We define for each $\varepsilon > 0$ Hilbert space $U_\varepsilon := H^2(0, L/\varepsilon) \cap H_0^1(0, L/\varepsilon)$ equipped with an inner product

$$(h, w)_{U_\varepsilon} := \int_{0}^{L/\varepsilon} \left(h'' w'' + \varepsilon^4 \, h \, w \right) dx. \tag{5.95}$$

Note that two norms that are induced by inner products of U_ε and V_ε are equivalent in $H^2(0, L/\varepsilon) \cap H_0^1(0, L/\varepsilon)$, more precisely for all sufficiently small $\varepsilon > 0$ one has

$$\sqrt{\frac{2}{3}} \varepsilon^2 \|h\|_{V_\varepsilon} \leq \|h\|_{U_\varepsilon} \leq \|h\|_{V_\varepsilon}. \tag{5.96}$$

The second inequality in (5.96) is trivial. Let us show the first one. Using integration by parts one obtains

$$\|h\|_{V_\varepsilon}^2 = \sum_{k=0}^{2} \int_{0}^{L/\varepsilon} (h^{(k)})^2 \, dx =$$

$$= \int_{0}^{L/\varepsilon} (h'')^2 \, dx - \int_{0}^{L/\varepsilon} h'' h \, dx + \int_{0}^{L/\varepsilon} h^2 \, dx$$

$$\leq \frac{3}{2} \left(\int_{0}^{L/\varepsilon} (h'')^2 \, dx + \int_{0}^{L/\varepsilon} h^2 \, dx \right).$$

From this and (5.95) one obtains the first inequality in (5.96).

Proposition 5.32. *For each $\varepsilon > 0$ there exist operators $A_\varepsilon, B_\varepsilon \in \mathcal{L}(U_\varepsilon, V_\varepsilon)$ such that*

$$(A_\varepsilon h, w)_{V_\varepsilon} := g_\varepsilon(h, w), \ (B_\varepsilon h, w)_{V_\varepsilon} := (h, w)_{H_\varepsilon},$$

for all $h, w \in H^2(0, L/\varepsilon) \cap H_0^1(0, L/\varepsilon)$.

Proof: Using definitions (5.94), (5.26), (5.24a)–(5.24b) and Cauchy-Schwarz inequality one obtains

$$|g_\varepsilon(h, w)| \leq \varepsilon^{-2} \|h\|_{U_\varepsilon} \|w\|_{V_\varepsilon} \text{ and } |(h, w)_{H_\varepsilon}| \leq \varepsilon^{-2} \|h\|_{U_\varepsilon} \|w\|_{V_\varepsilon}. \tag{5.97}$$

Therefore, for any $\varepsilon > 0$ and $h \in U_\varepsilon$ functionals $g_\varepsilon(h, \cdot)$ and $b_\varepsilon(h, \cdot)$ defined on V_ε are linear and continuous. Next, from the Riesz theorem it follows that there exists $v_\varepsilon(h), z_\varepsilon(h) \in V_\varepsilon$ such that

$$(v_\varepsilon(h), w)_{V_\varepsilon} = g_\varepsilon(h, w), \ (z_\varepsilon(h), w)_{V_\varepsilon} = (h, w)_{H_\varepsilon}.$$

Define then $B_\varepsilon h := z_\varepsilon(h)$ and $A_\varepsilon h := v_\varepsilon(h)$. Operators $A_\varepsilon, B_\varepsilon$ are linear and from (5.97) it follows that

$$\|A_\varepsilon\|_{\mathcal{L}(U_\varepsilon, V_\varepsilon)} \leq \varepsilon^{-2} \text{ and } \|B_\varepsilon\|_{\mathcal{L}(U_\varepsilon, V_\varepsilon)} \leq \varepsilon^{-2}.$$

∎

5.6 Existence of Eigenvalues with Prescribed Asymptotics

Note that for all $h, w \in H^2(0, L/\varepsilon) \cap H^1_0(0, L/\varepsilon)$ one has

$$(A_\varepsilon h, w)_{V_\varepsilon} = (h, A_\varepsilon w)_{V_\varepsilon} \text{ and } (B_\varepsilon h, w)_{V_\varepsilon} = (h, B_\varepsilon w)_{V_\varepsilon}.$$

Now, we can write EVP problem "from above" as a generalized EVP

$$h \in U_\varepsilon, \lambda \in \mathbb{R} : A_\varepsilon h = \lambda B_\varepsilon h.$$

The set of solutions $[h^j_\varepsilon, \lambda^j(\varepsilon)]$ of this EVP coincides with that one of (5.27) with $i = 2$. Motivated by Remark 5.28 let us define for each $j \in \mathbb{N}_0$ approximations for $[h^j_\varepsilon, \lambda^j(\varepsilon)]$ as

$$H^j_\varepsilon(x) = C_\varepsilon \begin{cases} \varepsilon x - \dfrac{\varepsilon^3}{3a^2_\varepsilon}x^3 + p^j_\varepsilon(x), & x \leq a_\varepsilon/\varepsilon \\ \dfrac{2a_\varepsilon}{3}\sin\left(\sqrt{K^j}(\varepsilon x - L)\right), & x \geq a_\varepsilon/\varepsilon \end{cases}, \quad \Lambda^j_\varepsilon = K^j \varepsilon^2, \quad (5.98)$$

where

$$K^j := \left(\frac{\pi(2j+1)}{2(L-b_\varepsilon)}\right)^2.$$

Constant C_ε in (5.98) is chosen to fulfill a normalization condition

$$\|H^j_\varepsilon\|_{H_\varepsilon} \equiv 1 \text{ for all } \varepsilon > 0.$$

Using definitions (5.98) and (5.26) it is easy to check that

$$C_\varepsilon = O(\varepsilon^{1/2}). \quad (5.99)$$

The polynomial correction term $p^j_\varepsilon(x) = O(\varepsilon^s)$, $s > 0$ for all $x \in [0, a_\varepsilon/\varepsilon]$ and $j \in \mathbb{N}_0$ is chosen so that $H^j_\varepsilon \in V_\varepsilon$ (in particularly $H^j_\varepsilon(x)$ is continuously differentiable on $[0, L/\varepsilon]$) and to provide the control on the first three derivatives of $H^j_\varepsilon(x)$ at the point $x = a/\varepsilon$:

$$\left|\frac{dH^j_\varepsilon}{dx}(a/\varepsilon)\right| + \sum_{k=2}^{3}\left|\frac{d^{(k)}H^j_\varepsilon}{dx^{(k)}}(a/\varepsilon - 0) - \frac{d^{(k)}H^j_\varepsilon}{dx^{(k)}}(a/\varepsilon + 0)\right| \leq \text{const } \varepsilon^{4+s}, \quad k = 2, 3. \quad (5.100)$$

Next, take the Banach spaces in the formulation of Theorem 5.30 as

$$Y_\varepsilon := \mathbb{R} \times U_\varepsilon \text{ and } Z_\varepsilon := \mathbb{R} \times V_\varepsilon$$

with the following norms:

$$\|(z, \mu)\|_{Y_\varepsilon} := \|z\|_{U_\varepsilon} + |\mu|, \|(z, \mu)\|_{Z_\varepsilon} := \|z\|_{V_\varepsilon} + |\mu|,$$

and define for each $j \in \mathbb{N}_0$ an operator function

$$\overline{F}^j_\varepsilon : Y_\varepsilon \to Z_\varepsilon, \quad \overline{F}^j_\varepsilon\left(\overline{h}, \overline{\lambda}\right) := \frac{1}{\varepsilon^2}\begin{bmatrix} A_\varepsilon \overline{h} - \overline{\lambda} B_\varepsilon \overline{h} \\ \left(\overline{h}, H^j_\varepsilon\right)_{H_\varepsilon} - 1 \end{bmatrix}. \quad (5.101)$$

Finally, the operator implicit function from Theorem 5.30 corresponding to the sought eigenpair $[h^j_\varepsilon, \lambda^j(\varepsilon)]$ is defined by an appropriate shift of the variables in (5.101)

$$F^j_\varepsilon(h, \lambda) = \overline{F}^j_\varepsilon\left(h + H^j_\varepsilon, \varepsilon^2\lambda + \Lambda^j_\varepsilon\right). \quad (5.102)$$

Chapter 5 Spectrum Asymptotics in a Singular Limit

Step2–proof of the assumption (5.88):

We show here that for sufficiently small $\varepsilon > 0$

$$||F_\varepsilon^j(0,0)||_{Z_\varepsilon} \leq \text{const}\, \varepsilon^{\nu_1} \quad \text{with } \nu_1 := 2+s. \tag{5.103}$$

Using (5.102) and $||H_\varepsilon^j||_{H_\varepsilon} \equiv 1$ one obtains

$$||F_\varepsilon^j(0,0)||_{Z_\varepsilon} = ||\overline{F}_\varepsilon^j(H_\varepsilon^j, \Lambda_\varepsilon^j)||_{Z_\varepsilon} = \sup_{||w||_{V_\varepsilon}=1} \left| \frac{1}{\varepsilon^2}\left(A_\varepsilon H_\varepsilon^j - \Lambda_\varepsilon^j B_\varepsilon H_\varepsilon^j, w\right)_{V_\varepsilon} \right|$$

$$= \sup_{||w||_{V_\varepsilon}=1} \frac{1}{\varepsilon^2} \left| g_\varepsilon(H_\varepsilon^j, w) - \Lambda_\varepsilon^j(H_\varepsilon^j, w)_{H_\varepsilon} \right|$$

Let us show

$$\sup_{||w||_{V_\varepsilon}=1} \left| g_\varepsilon(H_\varepsilon^j, w) - \Lambda_\varepsilon^j(H_\varepsilon^j, w)_{H_\varepsilon} \right| \leq \text{const}\, \varepsilon^{4+s}. \tag{5.104}$$

Using definitions (5.24a)–(5.24b) and two times integration by parts one obtains

$$\left| g_\varepsilon(H_\varepsilon^j, w) - \Lambda_\varepsilon^j(H_\varepsilon^j, w)_{H_\varepsilon} \right| = \left| \int_0^{a_\varepsilon/\varepsilon} \left(\frac{d^4 H_\varepsilon^j}{dx^4} - \varepsilon^4 \Lambda_\varepsilon^j H_\varepsilon^j \right) w\, dx \right.$$

$$+ \int_{a_\varepsilon/\varepsilon}^{L/\varepsilon} \left(\frac{d^4 H_\varepsilon^j}{dx^4} - \frac{d^2 H_\varepsilon^j}{dx^2} - \Lambda_\varepsilon^j(1-\varepsilon^{1/12})H_\varepsilon^j \right) w\, dx$$

$$+ \left(\frac{d^2 H_\varepsilon^j}{dx^2}(a_\varepsilon/\varepsilon - 0) - \frac{d^2 H_\varepsilon^j}{dx^2}(a_\varepsilon/\varepsilon + 0) \right) w'(a_\varepsilon/\varepsilon)$$

$$\left. - \left(\frac{d^3 H_\varepsilon^j}{dx^3}(a_\varepsilon/\varepsilon - 0) - \frac{d^3 H_\varepsilon^j}{dx^3}(a_\varepsilon/\varepsilon + 0) + \frac{dH_\varepsilon^j}{dx}(a) \right) w(a_\varepsilon/\varepsilon) \right|. \tag{5.105}$$

Let us estimate every term in the last expression separately. By (5.98), (5.99) and Cauchy-Schwarz inequality one has:

$$\left| \int_0^{a_\varepsilon/\varepsilon} \frac{d^4 H_\varepsilon^j}{dx^4} w\, dx \right| \leq \text{const}\, \varepsilon^{4+s}||w||_{V_\varepsilon}$$

$$\left| \int_0^{a_\varepsilon/\varepsilon} \varepsilon^4 \Lambda_\varepsilon^j H_\varepsilon^j w\, dx \right| \leq \text{const}\, \varepsilon^6 ||w||_{V_\varepsilon}.$$

Analogously

$$\left| \frac{d^4 H_\varepsilon^j}{dx^4} - \frac{d^2 H_\varepsilon^j}{dx^2} - \Lambda_\varepsilon^j(1-\varepsilon^{1/12})H_\varepsilon^j \right| \leq \text{const}\, \varepsilon^{4+s} \text{ for all } x \in (a_\varepsilon/\varepsilon, L/\varepsilon],$$

by the direct substitution of (5.98), (5.99) in it. Finally, using Cauchy-Schwarz inequality and the fact that $||w||_{V_\varepsilon} = 1$ one can show that

$$|w'(a_\varepsilon/\varepsilon) + w(a_\varepsilon/\varepsilon)| \leq ||w||_{V_\varepsilon} = 1.$$

Applying last three estimates and (5.100) to expression (5.105) one arrives at (5.104). The latter in turn implies estimate (5.103). Therefore, assumption (5.88) is shown with $\nu_1 = 2+s$.

5.6 Existence of Eigenvalues with Prescribed Asymptotics

Step3–proof of the assumption (5.89):
The differential of the function $F_\varepsilon^j(\lambda, h)$ is of the form,

$$(F_\varepsilon^j)'(h, \lambda) \begin{bmatrix} z \\ \mu \end{bmatrix} = \frac{1}{\varepsilon^2} \begin{bmatrix} A_\varepsilon z - \lambda\varepsilon^2 B_\varepsilon z - \Lambda_\varepsilon^j B_\varepsilon z - \mu\varepsilon^2 B_\varepsilon (h + H_\varepsilon^j) \\ (z, H_\varepsilon^j)_{H_\varepsilon} \end{bmatrix}.$$

From this it follows that

$$\left[(F_\varepsilon^j)'(h, \lambda) - (F_\varepsilon^j)'(0, 0)\right] \begin{bmatrix} z \\ \mu \end{bmatrix} = \begin{bmatrix} -\lambda B_\varepsilon z - \mu B_\varepsilon h \\ 0 \end{bmatrix}.$$

Assumption (5.89) is satisfied with $\nu_2 := 2$ if one proves that there exists a constant $K_2 > 0$ such that for all $z \in U_\varepsilon$, $\mu \in \mathbb{R}$

$$||\lambda B_\varepsilon z + \mu B_\varepsilon h||_{V_\varepsilon} \leq K_2 \frac{||h||_{U_\varepsilon} + |\lambda|}{\varepsilon^2} (||z||_{U_\varepsilon} + |\mu|). \tag{5.106}$$

This follows from the following estimate:

$$||\lambda B_\varepsilon z + \mu B_\varepsilon h||_{V_\varepsilon} = \sup_{||v||_{V_\varepsilon}=1} \left|\left(B_\varepsilon(\lambda z + \mu h), v\right)_{V_\varepsilon}\right|$$

$$= \sup_{||v||_{V_\varepsilon}=1} \left|\int_0^{L/\varepsilon} f_\varepsilon^2(x)(\lambda z + \mu h)v \, dx\right|$$

$$\leq ||f_\varepsilon^2||_\infty \sup_{||v||_{V_\varepsilon}=1} \int_0^{L/\varepsilon} |\lambda z + \mu h||v| \, dx \leq$$

$$\leq ||\lambda z + \mu h||_{V_\varepsilon} \leq \frac{\max\{|\lambda|, ||h||_{U_\varepsilon}\}}{\sqrt{2/3}\,\varepsilon^2}(||z||_{U_\varepsilon} + |\mu|),$$

where we used Cauchy-Schwarz inequality, definition (5.24b) and estimate (5.96). Assumption (5.89) is proved with $\nu_2 = 2$.

Step4–proof of the assumption (5.90):
One can prove that $(F_\varepsilon^j)'(0, 0)$ is Fredholm of index zero by decomposing it into the sum of bijective operator plus the rest and then applying Relich-Kondrachov compactness theorem (Theorem 6.2 of Adams [51]) in order to prove that the latter one is compact. Let us prove that there are no sequences $\varepsilon_n > 0$, $z_n \in U_{\varepsilon_n}$ and $\mu_n \in \mathbb{R}$ such that

$$||z_n||_{U_{\varepsilon_n}} + |\mu_n| = 1 \tag{5.107}$$

and

$$|\varepsilon_n| + \left\|\left(F_{\varepsilon_n}^j\right)'(0,0) \begin{bmatrix} z_n \\ \mu_n \end{bmatrix}\right\|_{Z_{\varepsilon_n}} \to 0 \text{ as } n \to \infty.$$

The last expression is equivalent to

$$\begin{cases} |\varepsilon_n| + \left\|\frac{1}{\varepsilon_n^2}A_{\varepsilon_n}z_n - B_{\varepsilon_n}\left(K^j z_n + \mu_n H_{\varepsilon_n}^j\right)\right\|_{V_{\varepsilon_n}} \to 0 \\ \frac{1}{\varepsilon_n^2}(z_n, H_{\varepsilon_n}^j)_{H_{\varepsilon_n}} \to 0 \end{cases}.$$

Let us use a contradiction argument. Suppose that there exist sequences ε_n, z_n and μ_n that

Chapter 5 Spectrum Asymptotics in a Singular Limit

satisfy expressions above. Then the last expression yields

$$\sup_{\|w\|_{V_{\varepsilon_n}}=1}\left|\left(\frac{1}{\varepsilon_n^2}A_{\varepsilon_n}z_n - B_{\varepsilon_n}\left(K^j z_n + \mu_n H_{\varepsilon_n}^j\right), w\right)\right|_{V_{\varepsilon_n}} \to 0 \text{ as } n \to \infty. \tag{5.108}$$

a) In (5.108) take a sequence

$$w_n := \frac{H_{\varepsilon_n}^m}{\|H_{\varepsilon_n}^m\|_{V_{\varepsilon_n}}}$$

corresponding to a fixed $m \in \mathbb{N}_0$. By (5.98) and (5.99) for all $m \in \mathbb{N}_0$ there is a constant $K > 0$ such that

$$\|H_{\varepsilon_n}^m\|_{V_{\varepsilon_n}} \leq K$$

From this and (5.108) it follows

$$\left(\frac{1}{\varepsilon_n^2}A_{\varepsilon_n}z_n - B_{\varepsilon_n}\left(K^j z_n + \mu_n H_{\varepsilon_n}^j\right), H_{\varepsilon_n}^m\right)_{V_{\varepsilon_n}} \to 0 \text{ for all } m \in \mathbb{N}_0. \tag{5.109}$$

Taking $m = j$ in the last expression, using definitions of operators A_{ε_n}, B_{ε_n} and $\|H_{\varepsilon_n}^j\|_{H_{\varepsilon_n}} \equiv 1$ one obtains

$$\left(z_n, \frac{1}{\varepsilon_n^2}A_{\varepsilon_n}H_{\varepsilon_n}^m - K^m B_{\varepsilon_n}H_{\varepsilon_n}^m\right)_{V_{\varepsilon_n}} - \mu_n \to 0.$$

Let us estimate the first term in the last expression

$$\left(z_n, \frac{1}{\varepsilon_n^2}A_{\varepsilon_n}H_{\varepsilon_n}^m - K^m B_{\varepsilon_n}H_{\varepsilon_n}^m\right)_{V_{\varepsilon_n}} \leq \|z\|_{V_{\varepsilon_n}} \|\frac{1}{\varepsilon_n^2}A_{\varepsilon_n}H_{\varepsilon_n}^m - K^m B_{\varepsilon_n}H_{\varepsilon_n}^m\|_{V_{\varepsilon_n}}$$

$$\leq \text{const}\, \frac{\|z\|_{U_{\varepsilon_n}}}{\varepsilon_n^2}\varepsilon_n^{2+s} \leq \text{const}\, \varepsilon_n^s \to 0 \text{ as } n \to \infty, \tag{5.110}$$

where we used (5.107), estimate (5.96) and approximation property (5.103). Therefore, it follows that

$$\mu_n \to 0 \text{ as } n \to \infty. \tag{5.111}$$

Consequently by (5.107) one obtains

$$\|z_n\|_{U_{\varepsilon_n}} \to 1 \text{ as } n \to \infty. \tag{5.112}$$

Taking now $m \neq j$ in (5.109) and using again (5.110) one obtains

$$\left|\left(K^m - K^j\right)\left(z_n, H_{\varepsilon_n}^m\right)_{H_{\varepsilon_n}} - \mu_n \left(H_{\varepsilon_n}^j, H_{\varepsilon_n}^m\right)_{H_{\varepsilon_n}} + O(\varepsilon_n^s)\right| \to 0. \tag{5.113}$$

Using definitions (5.98), (5.26), (5.24b) one gets:

$$\left(H_{\varepsilon_n}^j, H_{\varepsilon_n}^m\right)_{H_{\varepsilon_n}} = \int_0^{b_{\varepsilon_n}/\varepsilon_n} H_{\varepsilon_n}^j H_{\varepsilon_n}^m \varepsilon^4\, dx + \int_{b_{\varepsilon_n}/\varepsilon_n}^{L/\varepsilon_n} \left(1 - \varepsilon^{1/12}\right) H_{\varepsilon_n}^j H_{\varepsilon_n}^m\, dx =$$

$$= \int_0^{b_{\varepsilon_n}/\varepsilon_n} H_{\varepsilon_n}^j H_{\varepsilon_n}^m \varepsilon^4\, dx.$$

From (5.98) and (5.99) it follows that

$$\int_0^{b_{\varepsilon_n}/\varepsilon_n} H_{\varepsilon_n}^j H_{\varepsilon_n}^m\, dx \leq \text{const for all } n \in \mathbb{N} \text{ uniformly in } m,$$

and therefore one has
$$\left(H_{\varepsilon_n}^j, H_{\varepsilon_n}^m\right)_{H_{\varepsilon_n}} \to 0 \text{ uniformly in } m. \tag{5.114}$$

Using this and (5.111) one obtains from (5.113) that
$$\left|(K^m - K^j)(z_n, H_{\varepsilon_n}^m)_{H_{\varepsilon_n}}\right| \to 0 \text{ uniformly in } m.$$

Further by (5.96) and (5.107) one obtains
$$\left|\int_0^{b_{\varepsilon_n}/\varepsilon_n} \varepsilon^4 z_n H_{\varepsilon_n}^m \, dx\right| \leq \sqrt{3/2}\varepsilon^2 \|z_n\|_{U_{\varepsilon_n}} \sqrt{\int_0^{b_{\varepsilon_n}/\varepsilon_n} H_{\varepsilon_n}^m H_{\varepsilon_n}^m \, dx} \to 0.$$

From this and (5.115) using again definitions (5.26), (5.24b) one arrives at
$$(K^m - K^j) \int_{b_{\varepsilon_n}/\varepsilon_n}^{L/\varepsilon_n} z_n H_{\varepsilon_n}^m \, dx \to 0 \text{ uniformly in } m \in \mathbb{N}_0. \tag{5.115}$$

Using definition (5.98) and the fact that trigonometric system $\sin((2m+1)x)$ for $m \in \mathbb{N}_0$ forms an orthogonal basis in $L^2(0, \pi/2)$ one can show that the system $\{H_{\varepsilon_n}^m(x)\}$, $m \in \mathbb{N}_0$ restricted to $(b_{\varepsilon_n}/\varepsilon_n, L/\varepsilon_n)$ forms an orthogonal basis in $L^2(b_{\varepsilon_n}/\varepsilon_n, L/\varepsilon_n)$ for all $n \in \mathbb{N}$. One can also show that there exist a constant $\tilde{K} > 0$ such that
$$K_m := \sqrt{\int_{b_{\varepsilon_n}/\varepsilon_n}^{L/\varepsilon_n} (H_{\varepsilon_n}^m)^2 \, dx} \geq \tilde{K} \text{ for all } m \in \mathbb{N}_0.$$

Next, using Parseval's identity for the orthonormal basis formed by the restriction of the system $\{H_{\varepsilon_n}^m(x)/K_m\}$ to the interval $(b_{\varepsilon_n}/\varepsilon_n, L/\varepsilon_n)$ and (5.115) one obtains that
$$\int_{b_{\varepsilon_n}/\varepsilon_n}^{L/\varepsilon_n} z_n^2 \, dx \to 0 \text{ as } n \to \infty.$$

From this using for the third time definitions (5.26), (5.24b) one obtains
$$(z_n, z_n)_{H_{\varepsilon_n}} \to 0 \text{ as } n \to \infty. \tag{5.116}$$

b) Take in (5.108) a sequence of test functions $w_n := \varepsilon^2 z_n$ with $\|w_n\|_{V_\varepsilon} \leq 1$ by (5.96). Then by (5.111), (5.114), (5.116) it follows,
$$(A_{\varepsilon_n} z_n, z_n)_{V_{\varepsilon_n}} \to 0 \text{ as } n \to \infty.$$

In particularly, using definition of operator A_{ε_n} one gets
$$\int_0^{L/\varepsilon_n} (z_n'')^2 \, dx \to 0 \text{ as } n \to \infty. \tag{5.117}$$

c) We are going to show $\varepsilon_n^4 \|z_n\|_{L^2(0, L/\varepsilon_n)}^2 \to 0$ as $n \to \infty$. Define functions $p_n(\bar{x}) := \varepsilon_n^{3/2} z_n(\bar{x}/\varepsilon_n)$, where $\bar{x} := \varepsilon_n x$. Then for every $n \in \mathbb{N}$ it follows that $p_n(\bar{x}) \in H^2(0, L) \cap H_0^1(0, L)$. Moreover, by (5.117) it follows
$$\int_0^L (p_n''(\bar{x}))^2 d\bar{x} = \int_0^{L/\varepsilon_n} (z_n''(x))^2 \, dx \to 0 \text{ as } n \to \infty.$$

Chapter 5 Spectrum Asymptotics in a Singular Limit

By the fact that Laplacian forms an isomorphism between $L^2(0, L)$ and $H^2(0, L) \cap H_0^1(0, L)$ (see Theorem 8.12 of Gilbard and S.Trudinger [52]) it follows that

$$\int_0^L (p_n(\bar{x}))^2 \, d\bar{x} \to 0 \text{ as } n \to \infty$$

and hence using definition of $p_n(\bar{x})$ one obtains

$$\varepsilon^4 \int_0^{L/\varepsilon_n} (z_n(x))^2 \, dx \to 0 \text{ as } n \to \infty. \tag{5.118}$$

Finally, combining this and (5.117) and using definition (5.95) we conclude that $\|z_n\|_{U_{\varepsilon_n}} \to 0$ as $n \to \infty$. But this contradicts to (5.112) and hence the assumption of Lemma 5.31 holds. Then Lemma 5.31 implies the assumption (5.90) and we conclude that three assumptions of Theorem 5.30 hold applied to function (5.102).

Step5–results of the modified implicit function theorem:
Theorem 5.30 implies that for every $j \in \mathbb{N}_0$ there exist $\varepsilon^j > 0$ and functions $[h_\varepsilon^j, \lambda^j(\varepsilon)]$ such that

$$\overline{F}_\varepsilon^j(h_\varepsilon^j, \lambda^j(\varepsilon)) \equiv 0 \text{ for all } \varepsilon \in (0, \varepsilon^j),$$

where function $\overline{F}_\varepsilon^j$ is defined in (5.101). Therefore $[h_\varepsilon^j, \lambda^j(\varepsilon)]$ is a solution of the approximate EVP "from above". Moreover, by (5.91) and the approximation property (5.103) it follows that there exists a constant $c^j > 0$ such that for all $\varepsilon \in (0, \varepsilon^j)$

$$\|h_\varepsilon^j - H_\varepsilon^j\|_{U_\varepsilon} \leq c^j \varepsilon^{2+s}, \quad |\lambda^j(\varepsilon) - \Lambda_\varepsilon^j| \leq c^j \varepsilon^{4+s}.$$

Therefore, function $\lambda^j(\varepsilon)$ satisfies both assertions of the theorem. ∎

5.7 Proof of the Main Theorems

In this section we prove Theorems 5.10–5.12.
Proof of Theorem 5.12: We proceed analogously to the proof of Theorem 5.29 based on application of the modified implicit function Theorem 5.30. We transform first EVP (5.18) to an operator form. Define operators $A_\varepsilon \in \mathcal{L}(W_\varepsilon)$ and $B_\varepsilon \in \mathcal{L}(W_\varepsilon)$ by

$$(A_\varepsilon h, w)_{W_\varepsilon} := \int_{-L/\varepsilon}^{L/\varepsilon} h'' w'' - r_\varepsilon h' w' \, dx,$$

$$(B_\varepsilon h, w)_{W_\varepsilon} := \int_{-L/\varepsilon}^{L/\varepsilon} h w f_\varepsilon \, dx \text{ for all } h, w \in W_\varepsilon,$$

where Hilbert space W_ε is defined in (5.16) and equipped with a standard $H^2(-L/\varepsilon, L/\varepsilon)$ inner product. Analogously to Proposition 5.32 one can show that for all $\varepsilon > 0$

$$\|A_\varepsilon\|_{\mathcal{L}(W_\varepsilon)} \leq 1 \text{ and } \|B_\varepsilon\|_{\mathcal{L}(W_\varepsilon)} \leq 1.$$

Next, we write (5.18) as a generalized EVP

$$h \in W_\varepsilon, \lambda \in \mathbb{R} : A_\varepsilon h = \lambda B_\varepsilon h.$$

Now for each $\varepsilon > 0$ consider a function $h_{0,\varepsilon}(x) - h_\varepsilon^-$, where $h_{0,\varepsilon}(x)$ is the stationary solution from Proposition 5.2, and h_ε^- is its minimum value attained at the points $x = \pm L/\varepsilon$ by (5.11c).

Motivated by observations stated before formulation of this theorem we define an approximation for the sought solution $[h_\varepsilon^*, \lambda^*(\varepsilon)]$ of EVP (5.18) as

$$H_\varepsilon^*(x) := C_\varepsilon(h_{0,\varepsilon}(x) - h_\varepsilon^-) \text{ and } \Lambda_\varepsilon^* := 0. \quad (5.119)$$

Again C_ε in (5.119) is chosen to fulfill a normalization condition

$$\|H_\varepsilon^*\|_\varepsilon := \sqrt{(H_\varepsilon^*, H_\varepsilon^*)_\varepsilon} \equiv 1 \text{ for all } \varepsilon > 0,$$

where we use a scalar product (5.17) in $L^2(-L/\varepsilon, L/\varepsilon)$. Using definitions (5.14b), (5.119) and asymptotics for $h_{0,\varepsilon}(x)$ given in Lemma 5.8 it is easy to check that again

$$C_\varepsilon = O(\varepsilon^{1/2}). \quad (5.120)$$

Further, $H_\varepsilon^*(x) \in C^\infty(-L/\varepsilon, L/\varepsilon)$. Similar to the proof of Theorem 5.29 define next an operator function

$$\overline{F}_\varepsilon^* : W_\varepsilon \times \mathbb{R} \to W_\varepsilon \times \mathbb{R}, \quad \overline{F}_\varepsilon^*\left(\overline{h}, \overline{\lambda}\right) := \frac{1}{\varepsilon^2} \left[\begin{array}{c} A_\varepsilon \overline{h} - \overline{\lambda} B_\varepsilon \overline{h} \\ \left(\overline{h}, H_\varepsilon^*\right)_\varepsilon - 1 \end{array} \right]. \quad (5.121)$$

Finally, the operator function to which we want to apply Theorem 5.30 is defined by

$$F_\varepsilon^*(h, \lambda) := \overline{F}_\varepsilon^*\left(h + H_\varepsilon^*, \varepsilon^2 \lambda\right).$$

The Banach spaces in the formulation of Theorem 5.30 we take both as

$$Y_\varepsilon = Z_\varepsilon := \mathbb{R} \times W_\varepsilon.$$

Let us now prove the first assumption of Theorem 5.30 for function F_ε^*. Below we show that there exists a constant $c > 0$ such that

$$\left| \int_{-L/\varepsilon}^{L/\varepsilon} \frac{d^2 H_\varepsilon^*}{dx^2} \frac{d^2 w}{dx^2} - r_\varepsilon \frac{dH_\varepsilon^*}{dx} \frac{dw}{dx} - \Lambda_\varepsilon^* f_\varepsilon H_\varepsilon^* w \, dx \right| \leq c\sqrt{\varepsilon} \exp\left(-\frac{\alpha}{\varepsilon^{2/3}}\right), \quad (5.122)$$

for all $w \in W_\varepsilon$ with $\|w\|_{W_\varepsilon} = 1$. Analogously to the proof of Theorem 5.29 this in turn implies approximation property (5.88) with any $\nu_1 > 0$, namely

$$\|F_\varepsilon^*(0,0)\|_{W_\varepsilon \times \mathbb{R}} \leq c\varepsilon^{-3/2} \exp\left(-\frac{\alpha}{\varepsilon^{2/3}}\right). \quad (5.123)$$

Let us then show (5.122). Applying two times integration by parts one obtains

$$\left| \int_{-L/\varepsilon}^{L/\varepsilon} \frac{d^2 H_\varepsilon^*}{dx^2} \frac{d^2 w}{dx^2} - r_\varepsilon \frac{dH_\varepsilon^*}{dx} \frac{dw}{dx} - \Lambda_\varepsilon^* f_\varepsilon H_\varepsilon^* w \, dx \right| =$$

$$= \left| \int_{-L/\varepsilon}^{L/\varepsilon} \left(\frac{d^4 H_\varepsilon^*}{dx^4} + \frac{d}{dx}\left(r_\varepsilon \frac{dH_\varepsilon^*}{dx}\right) - \Lambda_\varepsilon^* f_\varepsilon H_\varepsilon^* \right) w \, dx - \left(\frac{d^2 H_\varepsilon^*}{dx^2} \frac{dw}{dx} \right) \Big|_{-L/\varepsilon}^{L/\varepsilon} \right| =$$

$$= |C_\varepsilon h_{0,\varepsilon}''(L/\varepsilon)(w'(L/\varepsilon) - w'(-L/\varepsilon))| \leq c\sqrt{\varepsilon} \exp\left(-\frac{\alpha}{\varepsilon^{2/3}}\right),$$

where we also use definitions (5.119), (5.14a)–(5.14b), equation (5.11a) and estimates (5.47), (5.120). Therefore, for any $\nu_1 > 0$ there exists $\tilde{\varepsilon} > 0$ such that assumption (5.88) holds with ν_1 for all $\varepsilon \in (0, \tilde{\varepsilon})$.

Next, an analog to crucial estimate (5.106) in Theorem 5.29 for the proof of assumption (5.89)

Chapter 5 Spectrum Asymptotics in a Singular Limit

of Theorem 5.30 in this case has a form:

$$||\lambda B_\varepsilon z + \mu B_\varepsilon h||_{W_\varepsilon} \leq K_2(||h||_{W_\varepsilon} + |\lambda|)(||z||_{W_\varepsilon} + |\mu|).$$

Therefore, we conclude that assumption (5.89) holds with $\nu_2 := 0$. The proof of assumption (5.90) proceeds again using Lemma 5.31 and the contradiction argument as in the Step 4 of the proof of Theorem 5.29. We should just point out that in the current case linearized operator $(F_\varepsilon^*)'(0,0)$ has a more simple form

$$(F_\varepsilon^*)'(0,0)\begin{bmatrix} z \\ \mu \end{bmatrix} = \frac{1}{\varepsilon^2}\begin{bmatrix} A_\varepsilon z - \varepsilon^2 \mu B_\varepsilon H_\varepsilon^* \\ (z, H_\varepsilon^*)_{H_\varepsilon} \end{bmatrix},$$

and therefore in the points corresponding to **a)–b)** of Step 4 of the proof of Theorem 5.29 one needs to consider only two test sequences

$$w_n = \frac{H_{\varepsilon_n}^*}{||H_{\varepsilon_n}^*||_{W_{\varepsilon_n}}} \quad \text{and} \quad w_n = \frac{z_n}{||z_n||_{W_{\varepsilon_n}}}.$$

Finally, Theorem 5.30 implies that there exists $\varepsilon^* > 0$ and for each $\varepsilon \in (0, \varepsilon^*)$ functions $[h_\varepsilon^*, \lambda^*(\varepsilon)]$ such that

$$\overline{F}_\varepsilon^*(h_\varepsilon^*, \lambda^*(\varepsilon)) \equiv 0.$$

Therefore, $[h_\varepsilon^*, \lambda^*(\varepsilon)]$ is a solution to the symmetric EVP (5.18). Moreover, by (5.91) and approximation property (5.123) there exists a number $c^* > 0$ such that

$$||h_\varepsilon^* - H_\varepsilon^*||_{W_\varepsilon} \leq c^* \varepsilon^{-3/2} \exp\left(-\frac{\alpha}{\varepsilon^{2/3}}\right), \quad |\lambda^*(\varepsilon)| \leq c^* \varepsilon^{1/2} \exp\left(-\frac{\alpha}{\varepsilon^{2/3}}\right).$$

Theorem 5.30 also gives that there exists $\delta^* > 0$ such that for each $\varepsilon \in (0, \varepsilon^*)$ one has $\lambda^*(\varepsilon)$ is unique eigenvalue of EVP (5.18) in the band $|\lambda| \leq \delta^* \varepsilon^2$. The theorem is proved. ∎

Proof of Theorem 5.11: Let us first for each $j \in \mathbb{N}_0$ show the existence of eigenvalues $\lambda_D^j(\varepsilon)$ from the assertions of the theorem. By Theorem 5.24 and Theorem 5.29 there exists $\varepsilon^0 > 0$ such that for all $\varepsilon \in (0, \varepsilon^0)$

$$\lambda_\varepsilon^{2,0} = \left(\frac{\pi}{2(L - A/P)}\varepsilon\right)^2 + o(\varepsilon^2)$$

is the smallest eigenvalue of the EVP "from above" (i.e. of (5.27) for $i = 2$). Recall that the spectrum of (5.27) is ordered for each $\varepsilon > 0$ as stated in (5.28). From Proposition 5.20 and relation (5.33) it follows that for sufficiently small $\varepsilon > 0$

$$\lambda_\varepsilon^{2,0} \geq \lambda_{D,\varepsilon}^0 \geq \lambda_\varepsilon^{1,0},$$

where $\lambda_{D,\varepsilon}^0$ is the smallest eigenvalue of Dirichlet half-droplet EVP (5.21), which spectrum is ordered as stated in (5.31), and $\lambda_\varepsilon^{1,0}$ is the smallest eigenvalue of the EVP "from below" (i.e. of (5.27) for $i = 1$). From this and Theorem 5.24 applied to the EVP "from below" it follows that

$$\lambda_\varepsilon^{1,0} = \left(\frac{\pi}{2(L - A/P)}\varepsilon\right)^2 + o(\varepsilon^2),$$

and therefore also

$$\lambda_{D,\varepsilon}^0 = \left(\frac{\pi}{2(L - A/P)}\varepsilon\right)^2 + o(\varepsilon^2).$$

Next, let us consider eigenvalue $\lambda_\varepsilon^{2,1}$ from the ordering (5.28). Theorem 5.24, Theorem 5.29 and Lemma 5.27 yield that there exists $\varepsilon^1 > 0$ such that for all $\varepsilon \in (0, \varepsilon^1)$

$$\lambda_\varepsilon^{2,1} = \left(\frac{3\pi}{2(L - A/P)}\varepsilon\right)^2 + o(\varepsilon^2).$$

Using this and again relation (5.33) for $j = 1$, i.e.

$$\lambda_\varepsilon^{2,1} \geq \lambda_{D,\varepsilon}^1 \geq \lambda_\varepsilon^{1,1},$$

and applying Lemma 5.27 to eigenvalues $\lambda_\varepsilon^{1,0}$ and $\lambda_\varepsilon^{1,1}$ one obtains that also

$$\lambda_\varepsilon^{1,1} = \left(\frac{3\pi}{2(L - A/P)}\varepsilon\right)^2 + o(\varepsilon^2)$$

$$\lambda_{D,\varepsilon}^1 = \left(\frac{3\pi}{2(L - A/P)}\varepsilon\right)^2 + o(\varepsilon^2).$$

Proceeding further by induction one obtains that for every $j \in \mathbb{N}_0$ there exists a mapping $\lambda_D^j(\varepsilon)$ satisfying assertions (i) and (ii) of the theorem. Here we use that every eigenvalue of the Dirichlet half-droplet EVP (5.21) is also eigenvalue of the initial EVP (5.18).

Analogously, using Remarks 5.16, 5.25 and Theorem 5.12 one can show the existence of eigenvalues of the Neumann half-droplet problem (5.22)

$$\lambda_N^j(\varepsilon) = \left(\frac{\pi(2j+1)}{2(L - A/P)}\varepsilon\right)^2 + o(\varepsilon^2), \ j \in \mathbb{N}_0$$

and

$$\lambda^*(\varepsilon) = O\left(\varepsilon^{1/2} \exp\left(-\frac{\alpha}{\varepsilon^{2/3}}\right)\right).$$

Therefore, they are also eigenvalues of the initial EVP (5.18) and satisfy assertions (i) and (ii) of Theorem 5.11.

Finally, let us prove assertion (iii) of the theorem. From asymptotics

$$\lambda_D^j(\varepsilon) = \left(\frac{\pi(2j+1)}{2(L - A/P)}\varepsilon\right)^2 + o(\varepsilon^2)$$

and eigenvalue ordering (5.31) it follows that for every $j \in \mathbb{N}_0$ there exists $\delta^j > 0$ and $\varepsilon^j > 0$ such that for each $\varepsilon \in (0, \varepsilon^j)$ one has if λ is an eigenvalue of the Dirichlet half-droplet EVP with

$$\left|\lambda - \left(\frac{\pi(2j+1)}{2(L - A/P)}\varepsilon\right)^2\right| \leq \delta^j \varepsilon^2$$

then $\lambda = \lambda_{D,\varepsilon}^j$. The analogous rule, in general with different $\widetilde{\delta}^j > 0$, $\widetilde{\varepsilon}^j > 0$ for $j \in \{*, 0, 1, 2, ...\}$, should hold for eigenvalues of the Neumann half-droplet problem. Taking when necessary the minimum of δ^j, $\widetilde{\delta}^j$ and of ε^j, $\widetilde{\varepsilon}^j$ and using the fact that any eigenvalue of the EVP (5.18) is either eigenvalue of the Dirichlet (5.21) or Neumann (5.22) half-droplet EVPs implies assertion (iii) of the theorem. ∎

Proof of Theorem 5.10 is based on the proofs of Theorem 5.11–5.12 above. Following the former one for each $j \in \mathbb{N}_0$ function

$$\lambda_D^j(\varepsilon) = \left(\frac{\pi(2j+1)}{2(L-A/P)}\varepsilon\right)^2 + o(\varepsilon^2). \tag{5.124}$$

gives j-th eigenvalue of the Dirichlet half-droplet problem (5.21). Moreover, for sufficiently small $\varepsilon > 0$ there are no eigenvalues of (5.21) on the interval $(\lambda_{D,\varepsilon}^j, \lambda_{D,\varepsilon}^{j+1})$ and $\lambda_{D,\varepsilon}^0$ is the smallest one.

Take now a sequence $\{\varepsilon_l\} \to 0$ and let $\{\lambda_{\varepsilon_l}\}$ be a sequence of eigenvalues of the Dirichlet half-droplet EVP (5.21) such that there exists a positive number K^* with

$$\left|\frac{\lambda_{\varepsilon_l}}{\varepsilon_l^2}\right| \le K^* \text{ for all } l \in \mathbb{N}.$$

There exists $j \in \mathbb{N}$ such that

$$\left(\frac{\pi(2j+1)}{2(L-A/P)}\right)^2 \ge K^*$$

Therefore, by eigenvalue ordering (5.31) for all sufficiently large $l \in \mathbb{N}$ there exists $j_l \in \mathbb{N}_0$ such that $\lambda_{\varepsilon_l} = \lambda_{D,\varepsilon_l}^{j_l}$. Then asymptotics (5.124) implies that

$$\text{dist}\left(\frac{\lambda_{\varepsilon_l}}{\varepsilon_l^2}, M \setminus \{0\}\right) \to 0.$$

Completely analogously one can show that

$$\text{dist}\left(\frac{\lambda_{\varepsilon_l}}{\varepsilon_l^2}, M\right) \to 0$$

given a sequence $\{\varepsilon_l\} \to 0$ and a corresponding sequence $\{\lambda_{\varepsilon_l}\}$ of eigenvalues of the Neumann half-droplet problem (5.22) with $|\lambda_{\varepsilon_l}/\varepsilon_l^2| \le K^*$ for all $l \in \mathbb{N}$.

Now assertion (i) of the theorem follows from the fact that any eigenvalue of (5.18) is either eigenvalue of EVP (5.21) or one of EVP (5.22). The assertion (ii) of the theorem follows from the facts that

$$\lambda_{D,\varepsilon}^0, \lambda_{N,\varepsilon}^0 \sim \left(\frac{\pi}{2(L-A/P)}\varepsilon\right)^2$$

and that the smallest eigenvalue of EVP (5.18) λ_ε^* is negative, see Remark 5.6. ∎

5.8 Numerical Solutions and Comparison

Here we describe numerical solution of the EVP (5.12) and compare it with the leading order approximations (5.98), (5.119) for the set of eigenvalues of symmetric EVP (5.18), existence of which was proved in Theorems 5.11–5.12.

Our algorithm of numerical solution consists of three steps. Firstly, for fixed P, L and sufficiently small $\varepsilon > 0$ we solve (5.11a) with boundary conditions

$$h'_{0,\varepsilon}(\pm L/\varepsilon, P) = 0$$

numerically and calculate the stationary solution $h_{0,\varepsilon}(x)$. Using $h_{0,\varepsilon}(x)$ we then calculate the coefficient functions for the linear operator $\mathcal{L}_\varepsilon$. Secondly, we apply a finite difference discretization

5.8 Numerical Solutions and Comparison

j	0	1	2	3	4	5
$\lambda_{appr}/\varepsilon^2$	0.1381	1.2431	3.4532	6.7682	11.1883	16.7134
$\lambda_{num}/\varepsilon^2$	0.1437	1.2926	3.5843	7.0068	11.5425	17.1675

Table 5.1: Comparison of the first 6 eigenvalues for the Dirichlet eigenvalue problem (5.21) with $P = 0.1, L = 10, \epsilon = 10^{-6}$.

on a uniform mesh on the interval $[-L/\varepsilon, L/\varepsilon]$ to linear operator $\mathcal{L}_\varepsilon$ including also boundary conditions (5.9). The resulting approximation of our finite-difference scheme is $O(1/N^2)$, where N is a mesh size. Finally, the problem transforms to one of finding eigenvalues and eigenfunction of the matrix $A \in \mathbb{M}(N \times N)$ corresponding to discretized operator $\mathcal{L}_\varepsilon$. We calculate them using Implicitly Restarted Arnoldi Method, which was developed in Lehoucq and Sorensen [53] for the cases of large sparse matrices and implemented in the Fortran library ARPACK. The set of eigenpairs of matrix M give us a numerical approximation for the smallest eigenpairs of EVP (5.12). In Table 5.1 we compare first six eigenvalues calculated numerically (second row) for the Dirichlet half-droplet problem (5.21) and using analytical approximations (5.98) (first row). Similar agreement between numerical results and analytical approximations (5.98), (5.119) were obtained for eigenvalues of Neumann half-droplet problem (5.22). Our numerics also shows that for fixed $\varepsilon > 0$ and $j \in N_0$ the corresponding $\lambda_{D,\varepsilon}^j$ and $\lambda_{N,\varepsilon}^j$ are very close. In Figure 5.6 two numerically obtained eigenfunctions corresponding to eigenvalues $-\lambda_{D,\varepsilon}^2$ and $-\lambda_{N,\varepsilon}^2$ of initial EVP (5.12) are presented. As was stated in section 5.1 eigenfunctions of EVP (5.12) are derivatives of corresponding eigenfunctions of symmetric EVP (5.18). According to this and that the set of solutions to the latter problem is the union of solutions to Dirichlet and Neumann half-droplet EVPs (see section 5.3) Figure 5.6 shows that the left numerical eigenfunction is an even function and corresponds to the derivative of the eigenfunction for the Dirichlet half-droplet problem. The right one is an odd function and corresponds to the derivative of the eigenfunction for the Neumann half-droplet problem.

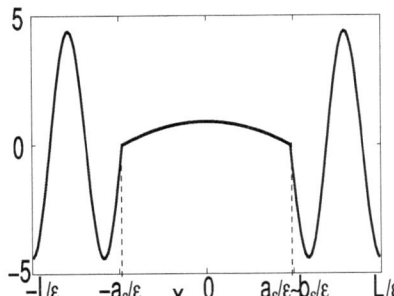
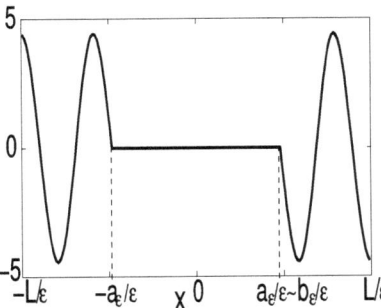

Figure 5.6: Eigenfunctions corresponding to eigenvalues $-\lambda_{D,\varepsilon}^2$ (left) and $-\lambda_{N,\varepsilon}^2$ (right) of EVP (5.12), $P = 0.1, L = 10, \epsilon = 10^{-6}$.

Figure 5.6 shows clearly two regions in the interval $[-L/\varepsilon, L/\varepsilon]$. In the outer interval both eigenfunctions are represented by trigonometric functions and in the inner one corresponding to the droplet core by polynomials. This stays in a good correspondence with the approximation for eigenfunctions from Remark 5.28 and (5.98). Due to chosen very small $\varepsilon = 10^{-6}$ and the fact that the relative length of the contact line interval between those regions tends to zero as

Chapter 5 Spectrum Asymptotics in a Singular Limit

$\varepsilon \to 0$ (see Lemma 5.8) numerically on Figure 5.6 we observe this interval as a point. Another observation concerns smoothness of the numerical eigenfunctions. In Figure 5.6 one can see that the first derivative of eigenfunctions is discontinuous at this point. This can be explained using the analytical result that we obtained for the approximate problem "from below" (see Remark 5.28), namely in the contact line region the derivative of eigenfunctions for the above problem oscillates fast proportionally to the negative power of ε. Therefore, it is a challenging numerical problem to resolve derivatives of eigenfunctions in this very small contact line region. Nevertheless numerical eigenfunctions from Figure 5.6 are continuous and this stays also in the correspondence with our analytical result (see Remark 5.28) which predicts that in the contact line region to the leading order in ε an eigenfunction itself is determined by a constant and do not possesses oscillations.

Finally, Figure 5.7 shows the numerical eigenfunction of EVP (5.12) corresponding to the exponentially small eigenvalue $-\lambda_\varepsilon^*$. Comparing Figures 5.5 and 5.7 one can see that this function is close to $h'_{0,\varepsilon}(x)$. This stays in a agreement with approximation (5.119) for the corresponding eigenfunction of the symmetric EVP (5.18).

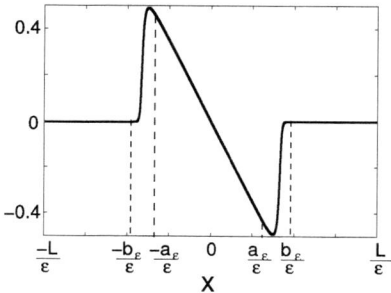

Figure 5.7: Eigenfunction corresponding to eigenvalue $-\lambda_\varepsilon^*$ of EVP (5.12), $P = 0.1$, $L = 20$, $\epsilon = 10^{-2}$.

Chapter 6
Summary and Outlook

In this study we considered the topic of derivation, analysis and numerics of reduced ODE models corresponding to lubrication equations (1.2), (1.3a)–(1.3b) describing physically weak-slip regime and strong-slip regime and their limiting cases (1.6), (1.8), (1.7a)–(1.7b) describing no-slip, intermediate and free suspended films regimes, respectively. The analysis of the reduced ODE models was geared towards the investigation of the influence of slippage on the coarsening dynamics of thin liquid polymer droplets. We summarize our results in detail in a list below.

- **Asymptotical derivation of reduced ODE models.** We derived asymptotically reduced ODE models (2.66) corresponding to the whole family of lubrication equations cited above. The reduced ODE models describe evolution in time for the set of pressures and positions of the droplets in an array. For each droplet in an array corresponds two equations, which are coupled with others through nonlinear functions corresponding to the mass fluxes that the droplet experiences due its neighbors. We found that the difference between the reduced model for general mobility model (1.9) which generalize no-, intermediate- and weak-slip cases, and the one for strong-slip model (1.3a)-(1.3b) lies in the mobility coefficients (2.28) and (2.42). Moreover, in the latter case the mobility coefficient essentially depends on the slip-length β.

- **Numerical investigation of reduced ODE models.** We proposed a numerical algorithm for the solution of the reduced systems, compared their results with those given by numerical solutions of initial lubrication equations and found a good agreement between them.

- **Influence of slippage on migration.** After derivation of the asymptotics for the mobility coefficient (2.42) of the reduced ODE model corresponding to the strong-slip equation (1.3a)–(1.3b) with respect to the small parameter ε, which appears in the pressure function (1.4), we found a unique critical slip-length β_{crit}, which decides the direction of migration of droplets. If the slip-length is smaller than above critical value the droplet migrates opposite to the direction of applied effective flux. If the slip-length is bigger than the critical value the droplet migrates in the direction of applied effective flux. We investigated then numerically the influence of this effect on the coarsening scenarios for arrays of many droplets and showed that the collision dominated coarsening rates depend on the value of slip-length.

Chapter 6 Summary and Outlook

- **Alternative derivation of reduced models using formal reduction onto an "approximate invariant" manifold.** Motivated by a center-manifold reduction approach of Mielke and Zelik [4] for semilinear parabolic equations we applied it formally for the derivation of the reduced ODE model for the case of the no-slip lubrication model. After that we compared the new reduced system with previously asymptotically derived one and found a good agreement. This formal approach could admit in future a rigorous justification via center-manifold reduction technique. We stated main open questions, which are needed to be solved for the rigorous justification of our formal approach and in the next point solved rigorously one of them.

- **Derivation of the spectral gap property.** We described asymptotics for the spectrum of the no-slip lubrication equation linearized at the stationary solution $h_{0,\varepsilon}$ introduced in Proposition 5.2 with respect to the small parameter ε tending to zero. This eigenvalue problem turned out to be a singular perturbed one. We showed rigorously the existence of an ε dependent spectral gap.

- **Existence of eigenvalues with prescribed asymptotics for singular perturbed eigenvalue problem.** Besides spectral gap property using a modified implicit function theorem first suggested by Recke and Omel'chenko [5] we showed the existence of eigenvalues with prescribed asymptotics, in particular of exponentially small one. These results offer a new technique for solving certain type of singular perturbed eigenvalue problems.

Finally, we would like to suggest three topics which could be considered as an extension of this study and interesting for an investigation in future.

- In section 2.5.2 we suggested a flux approximation in the reduced ODE model corresponding to strong-slip lubrication model (1.3a)–(1.3b). As we found there this approximation is valid only for relative small slip-lengths. In future we plan to derive an approximation which will satisfactory for all slip-lengths. This would allow us to simulate collision dominated coarsening rate with moderate β and put more light on analytical dependence of them on the slip-length.

- In section 2.7 we made preliminary numerical investigation of influence of inertia on the evolution governed by strong-slip lubrication model (1.3a)–(1.3b). They showed that for high Re numbers, which do not satisfy condition (2.29), the validity of reduced ODE models breaks down. Moreover, the height profile of the solution experiences oscillations. The possibility of new ODE reduced models which cover above mentioned effects is an interesting topic for future investigations.

- The spectral gap property shown in Chapter 5 indicates for a possibility of making the formal reduction approach of Chapter 4 rigorous by proving an existence of a center invariant manifold in a neighborhood of the 'approximate' one, at least in the case of one migrating droplet. This problem seems to be at the same rate interesting as difficult because it inherits the singular perturbed nature of linearized eigenvalue problem considered in Chapter 5. The next possible step in rigorous justification of it could be the proof of an analog of Theorem 8.5 of Mielke and Zelik [4], which states the existence of the center-manifold, using already known spectral gap property.

List of main symbols

Derivatives

$h'_\varepsilon(x)$, $h''_\varepsilon(x)$, $h^{(k)}_\varepsilon(x)$ derivatives w.r.t. x of function h with variables ε and x

∂_x, ∂_{xx}, ∂^k_x partial derivatives with respect to variable x

Constants

Re	Reynolds number
b, β	different scalings of the slip-length, page 4
A	droplet contact angle, page 64
k_1	maximum of function $r_\varepsilon(x)$, page 71
d, P_*, P^*	positive numbers introduced in Proposition 2.2, page 10

Functions and Operators

a_ε, b_ε	functions introduced in Lemma 5.8, page 67
$\beta_{crit}(\varepsilon, P)$	critical value of slippage, page 39
$E(h)$	Lyapunov functional (2.16) for (1.9), page 14
$E(u, h)$	Lyapunov functional (2.17) for (1.3a)–(1.3b), page 14
$\mathbb{F}_\varepsilon$	quasilinear elliptic operator (4.2), page 48
$M(h)$	mobility term in (1.9), page 5
$\mathbf{m}_\varepsilon$	diffeomorphism between $\mathbb{B}_\varepsilon$ and $\mathbb{P}_\varepsilon$, page 49
$\Pi(h)$	scaled intermolecular pressure function (5.8), page 65
$P_\mathbf{m}$	projection on the tangent space $\mathbb{T}_\mathbf{m}\mathbb{P}_\varepsilon$, page 56
$r_\varepsilon(x)$, $f_\varepsilon(x)$	coefficient functions in EVP (5.18), page 66
$r^1_\varepsilon(x)$, $f^1_\varepsilon(x)$	approximations from above for $r_\varepsilon(x)$, $f_\varepsilon(x)$, page 72
$r^2_\varepsilon(x)$, $f^2_\varepsilon(x)$	approximations from below for $r_\varepsilon(x)$, $f_\varepsilon(x)$, page 72
$T(\hat{h}_\varepsilon, \partial_x\hat{h}_\varepsilon, \partial_{xx}\hat{h}_\varepsilon)$	Trouton term (2.37), page 18
$U_\varepsilon(h)$	potential function (1.5), page 4
$U(h)$	scaled potential function (5.38), page 75
$\mathcal{U}_\varepsilon(h, P)$	function (2.7), page 10
$\mathcal{U}_\varepsilon(h)$	function (5.37a), page 75

List of main symbols

$\chi(x)$	characteristic function, page 49
$\chi_j(\mathbf{s})$, $j = 0, ..., N$	partitions of unity on interval $(-L, L)$, page 49
$\phi_j(\mathbf{s})$, $j = 0, ..., 2N - 1$	functions spanning the tangent space $\mathbb{T}_\mathbf{m}\mathbb{P}_\varepsilon$, page 49
$\psi_j(\mathbf{s})$, $j = 0, ..., 2N - 1$	'adjoint' functions, page 52
$\bar{\psi}_j(\mathbf{s})$, $j = 0, ..., 2N - 1$	functions defined in Proposition 4.4, page 55

Stationary solutions

$\hat{h}_\varepsilon(x, P)$	stationary solution to (1.9) on $\mathbb{R}$, page 9
$\hat{h}_\varepsilon^-(P)$, $\hat{h}_\varepsilon^+(P)$	minimum (2.3a) and maximum (2.3b) of $\hat{h}_\varepsilon(x, P)$, page 9
$\hat{h}_\varepsilon^c(P)$	elliptic center point to equation (2.4), page 9
$\overline{h}_{0,\varepsilon}(x, P)$	stationary solution to (5.1) with (5.2), page 64
$h_{0\varepsilon}(x)$	stationary solution to (5.7) with (5.9), page 65
h_ε^-, h_ε^+	minimum and maximum of $h_{0\varepsilon}(x)$, page 76
$\hat{h}_{sc,\varepsilon}(x)$	homoclinic solution to equation (5.34), page 75
$\hat{h}_{sc,\varepsilon}^-$, $\hat{h}_{sc,\varepsilon}^+$	minimum and maximum of $\hat{h}_{sc,\varepsilon}(x)$, page 75
$\hat{h}_{sc,\varepsilon}^c$	elliptic center point of equation (5.34), page 75

Sets and Spaces

$\mathbb{B}_\varepsilon$	open set (4.4) in $\mathbb{R}^{2N}$, page 48
M	discrete countable set (5.19), page 68
H_ε	$L^2(0, L/\varepsilon)$ with weighted inner product (5.26), page 72
$\mathbb{P}_\varepsilon$	'approximate invariant' manifold, page 49
$\partial \mathbb{P}_\varepsilon$	boundary of manifold $\mathbb{P}_\varepsilon$, page 49
σ_ε	spectrum of EVP (5.18) for a fixed $\varepsilon > 0$, page 68
$\mathbb{T}_\mathbf{m}\mathbb{P}_\varepsilon$	tangent space to manifold $\mathbb{P}_\varepsilon$ at a point $\mathbf{m}$, page 49
W_ε	$H^2(-L/\varepsilon, L/\varepsilon) \cap H_0^1(-L/\varepsilon, L/\varepsilon)$ with the standart inner product of $H^2(-L/\varepsilon, L/\varepsilon)$, page 66
V_ε	$H^2(0, L/\varepsilon) \cap H_0^1(0, L/\varepsilon)$ with the standart inner product of $H^2(0, L/\varepsilon)$, page 71
U_ε	$H^2(0, L/\varepsilon) \cap H_0^1(0, L/\varepsilon)$ with inner product (5.95), page 92
$(h, w)_\varepsilon$	weighted inner product (5.17) in $L^2(-L/\varepsilon, L/\varepsilon)$, page 66

Bibliography

[1] A. Münch, B. Wagner, and T. P. Witelski. Lubrication models with small to large slip lengths. *J. Engr. Math.*, 53:359–383, 2006.

[2] K. B. Glasner and T. P. Witelski. Coarsening dynamics of dewetting films. *Phys. Rev. E*, 67:016302, 2003.

[3] K. Glasner, F. Otto, T. Rump, and D. Slepjev. Ostwald ripening of droplets: the role of migration. *European J. Appl. Math.*, 20(1):1–67, 2009.

[4] A. Mielke and S. Zelik. Multi-pulse evolution and space-time chaos in dissipative systems. *Mem. Amer. Math. Soc.*, 198(925):1–97, 2009.

[5] L. Recke and O. Omel'chenko. Boundary layer solutions to singularly perturbed problems via the implicit function theorem. *J. Diff. Equations*, 245(12):3806–3822, 2008.

[6] A. Oron, S. H. Davis, and S. G. Bankoff. Long-scale evolution of thin liquid films. *Rev. Mod. Phys.*, 69(3):931–980, 1997.

[7] Steve Granick, Yingxi Zhu, and Hyunjung Lee. Slippery questions about complex fluids flowing past solids. *nature materials*, 2:221–227, 2003.

[8] K. Jacobs, S. Herminghaus, and H. Kuhlmann. Trendbericht mikrofluidik. Nachrichten aus der Chemie, 53:300–304, 2005.

[9] G. Reiter, A. Sharma, A. Casoli, M.-O. David, R. Khanna, and P. Auroy. Thin film instability induced by long range forces. *Langmuir*, 15:2551–2558, 1999.

[10] C. Redon, F. Brochard-Wyart, and F. Rondelez. Dynamics of dewetting. *Physical Review Letters*, 66(6):715–718, 1991.

[11] R Seemann, S. Herminghaus, and K. Jacobs. Gaining control of pattern formation of dewetting films. *Journal of Physics: Condensed Matter*, 13:4925–4938, 2001.

[12] P.G. de Gennes. Wetting: Statics and dynamics. *Review of Modern Physics*, 57:827, 1985.

[13] M. B. Williams and S. H. Davis. Nonlinear theory of film rupture. *J. Colloid Interface Sci.*, 90:220–228, 1982.

[14] T. Erneux and D. Gallez. Can repulsive forces lead to stable patterns in thin liquid films? *Phys. Fluids*, 9:1194–1196, 1997.

[15] A. L. Bertozzi, G. Grün, and T. P. Witelski. Dewetting films: bifurcations and concentrations. *Nonlinearity*, 14:1569–1592, 2001.

[16] A. Sharma and G. Reiter. Instability of thin polymer films on coated substrates: Rupture, dewetting and drop formation. J. Colloid Interface Sci., 178:383–389, 1996.

[17] F. Brochard-Wyart and C. Redon. Dynamics of liquid rim instabilities. *Langmuir*, 8:2324–2329, 1992.

Bibliography

[18] A. Münch and B. Wagner. Contact-line instability of dewetting thin films. *Physica D*, 209: 178–190, 2005.

[19] R. Limary and P. F. Green. Late-stage coarsening of an unstable structured liquid film. *Phys. Rev. E*, 60:021601, 2002.

[20] R. Limary and P. F. Green. Dynamics of droplets on the surface of a structured fluid film: Late-stage coarsening. *Langmuir*, 19:2419–2424, 2003.

[21] R. Fetzer, A. Münch, B. Wagner, M. Rauscher, and K. Jacobs. Quantifying hydrodynamic slip: A comprehensive analysis of dewetting profiles. *Langmuir*, 23:10559–10566, 2007.

[22] C. Neto, V. S. J. Craig, and D. R. M. Williams. Evidence of shear-dependent boundary slip in Newtonian liquids. Eur. Phys. J. E direct, page 10.1140/epjed/e2003, 2003.

[23] C. Redon, J. B. Brzoska, and F. Brochard-Wyart. Dewetting and slippage of microscopic polymer films. *Macromolecules*, 27:468–471, 1994.

[24] G. Reiter and A. Sharma. Auto-optimization of dewetting rates by rim instabilities in slipping polymer films. *PRL*, 80(16), 2001.

[25] R. Fetzer, K. Jacobs, A. Münch, B. Wagner, and T. P. Witelski. New slip regimes and the shape of dewetting thin liquid films. *Phys. Rev. Lett.*, 95:127801, 2005.

[26] K. Kargupta, A. Sharma, and R. Khanna. Instability, dynamics and morphology of thin slipping films. Langmuir, 20:244–253, 2004.

[27] M. P. Brenner and D. Gueyffier. On the bursting of viscous films. *Phys. Fluids*, 11(3): 737–739, 1999.

[28] D. Peschka. Self-similar rupture of thin liquid films with slippage. PhD Thesis, Institute of Mathematics, Humboldt University of Berlin, 2008.

[29] K. B. Glasner and T. P. Witelski. Collission vs. collapse of droplets in coarsening of dewetting thin films. *Physica D*, 209:80–104, 2005.

[30] J. W. Cahn and J. E. Hilliard. Free energy of a nonuniform system. I. Interfacial free energy. *J. Chem. Phys.*, 28:258–267, 1958.

[31] N. D. Alikakos, P. W. Bates, and G. Fusco. Slow motion for the Cahn–Hilliard equation in one space dimension. *J. Diff. Equations*, 90:81–135, 1991.

[32] P. W. Bates and J. P. Xun. Metastable patterns for the Cahn-Hilliard equation: Part I. *J. Diff. Equations*, 111:421–457, 1994.

[33] P. W. Bates and J. P. Xun. Metastable patterns for the Cahn-Hilliard equation: Part II. Layer dynamics and slow invariant manifold. *J. Diff. Equations*, 117:165–216, 1995.

[34] X. San and M. J. Ward. Dynamics and coarsening of interfaces for the viscous Cahn-Hilliard equation in one spatial dimension. *Studies in Applied Math*, 105:203–234, 2000.

[35] C. L. Emmott and A. J. Bray. Coarsening dynamics of a one-dmensional driven Cahn-Hilliard system. *Phys. Rev. E*, 54:4568–4575, 1996.

[36] S. J. Watson, F. Otto, B. Y. Rubinstein, and S. H. Davis. Coarsening dynamics of the convective Cahn-Hilliard equation. *J. Colloid Interface Sci.*, 22:127–148, 2003.

[37] F. Otto, T. Rump, and D. Slepjev. Coarsening rates for a droplet model: Rigorous upper bounds. *SIAM J. Appl. Math.*, 38:503–529, 2006.

[38] L. M. Pismen and Y. Pomeau. Mobility and interactions of weakly nonwetting droplets. *Phys. Fluids*, 16:2604–2612, 2004.

[39] K. B. Glasner. Ostwald ripening in thin film equations. *SIAM J. Appl. Math.*, 69:473–493, 2008.

[40] A. Münch. Dewetting rates of thin liquid films. *Journal of Physics: Condensed Matter*, 17: S309–S318, 2005.

[41] R. Magnus. The implicit function theorem and multibump solutions of periodic partial differential equations. Proc. Royal Soc. Edinburgh, 136:559–583, 2006.

[42] A. Erdelyi. *Asymptotic expansions*. Dover Publications, New York, 1956.

[43] D. Peschka, A. Münch, and B. Niethammer. Thin film rupture for large slip. submitted to J. Engr. Math, 2008.

[44] L. F. Shampine and M. K. Gordon. *Computer–Lösung gewöhnlicher Differentialgleichungen*. Friedr. Vieweg & Sohn Verlagsgesellschaft mbH, Braunschweig, 1984.

[45] H. Jeffreys and B. S. Jeffreys. *Methods of Mathematical Physics, 3rd ed.* Cambridge, England: Cambridge University Press, 1988.

[46] N. Fenichel. Geometric singular perturbation theory for ordinary differential equations. *J. Diff. Equations*, 31:53–98, 1979.

[47] C. Jones. Geometric singular perturbation theory. In L. Arnold, editor, *Dynamical Systems, Lecture Notes in Mathematics*, volume 1609, pages 44–118. Springer, Berlin, 1995.

[48] E. Kamke. *Gewöhnliche Differentialgleichungen*. Akademische Verlagsgesellschaft Geest&Portig K.-G., Liepzig, 1967.

[49] R. S. Laugesen and M. C. Pugh. Linear stability of steady states for thin film and Cahn-Hilliard type equations. Arch. Ration. Mech. Anal., 154:3–51, 2000.

[50] T. Kato. *Perturbation theory for linear operators*. Springer-Verlag, Berlin, 1995.

[51] R. A. Adams. *Sobolev Spaces*. Academic Press, New York-San Francisco-London, 1975.

[52] D. Gilbard and N. S.Trudinger. *Elliptic Partial Differential Equations of Second Order*. Springer Verlag, Berlin, 2001.

[53] R. B. Lehoucq and D. C. Sorensen. Deflation techniques for an implicitly re-started arnoldi iteration. *SIAM J. Matrix Analysis and Applications*, 17:789–821, 1996.